Weather, Religion and Climate Change

Weather, Religion and Climate Change is the first in-depth exploration of the fascinating way in which the weather impacts on the fields of religion, art, culture, history, science, and architecture.

In critical dialogue with meteorology and climate science, this book takes the reader beyond the limits of contemporary thinking about the Anthropocene and explores whether a deeper awareness of weather might impact on the relationship between nature and self. Drawing on a wide range of examples, including paintings by J.M.W. Turner, medieval sacred architecture, and Aristotle's classical *Meteorologica*, Bergmann examines a geographically and historically wide range of cultural practices, religious practices, and worldviews in which weather appears as a central, sacred force of life. He also examines the history of scientific meteorology and its ambivalent commodification today, as well as medieval "weather witchery" and biblical perceptions of weather as a kind of "barometer" of God's love. Overall, this volume explores the notion that a new awareness of weather and its atmospheres can serve as a deep cultural and spiritual driving force that can overcome the limits of the Anthropocene and open a new path to the "Ecocene", the age of nature.

Drawing on methodologies from religious studies, cultural studies, art history and architecture, philosophy, environmental ethics and aesthetics, history, and theology, this book will be of great interest to all those concerned with studying the environment from a transdisciplinary perspective on weather and wisdom.

Sigurd Bergmann's previous studies have investigated the relationship between the image of God and the view of nature in late antiquity and late modernity, the methodology of contextual theology, and visual arts in the indigenous Arctic and Australia, as well as visual arts, architecture and religion, and religion in climate change. He has initiated the *European Forum for the Study of Religion and the Environment*, and among his many publications are *Religion, Space and the Environment* (2014), *Religion in the Anthropocene* (ed. 2017), and *Arts, Religion and the Environment: Exploring Nature's Texture* (ed. 2018).

Taking inspiration from anthropologist Tim Ingold's "weather worlds", this fascinating book is a significant contribution to the environmental humanities that will be of interest to readers beyond the confines of theology and religious studies. Weaving in reflections on a rich collection of different art works, the author takes us inside pictorial, poetic, and symbolic narratives on what it means to live in our weather. By doing so, he not only challenges the habitual complacency of detachment in the West but also provokes new and innovative attitudes and practices in time and space. Rising to the surface like a deep subliminal current, Bergmann's theology contributes a spiritual dimension that recrafts the multilayered narratives that it interrogates.

Celia Deane-Drummond, Senior Research Fellow and Director of Laudato Si' Research Institute, Campion Hall, University of Oxford

Although we talk of the weather a great deal, we do not speak of the weather much in the study of religion. In this bold collection of essays, Bergmann invites us with his customary insight, generosity, and erudition into the liveliness of the weather. Refusing the greying out of the weather, Bergmann's analyses remind us that the weather is always colourfully intertwined with religious views and practices. This collection is a drenching gift and deserves to be read widely across the environmental humanities.

Peter Scott, The University of Manchester, UK

It was indeed difficult for me, while reading the book, to decide if Sigurd's book is a book of theology, history, science, or art! But soon I came to realise that it is in itself a piece of art. The reflections on contemporary environmental issues from a weather and theology perspective are impressive. Indeed, there is an essential need to bridge the gap between science and humanities, including religious values and beliefs, especially when we are facing an existential threat caused by our own consumptive behaviour. Reading the book gave me the inspiration of our need for a fourth industrial revolution, not based on technology, but on an ethical revolution that takes us back to our sacred values and beliefs. If we don't do so, soon not only Fiji will face climate migration, but even vast areas of coastal zones around the world. In many sacred books, God used weather to bring mercy to humans who faced social injustice, as Sigurd elaborates in the book, but in our days, it seems our unsustainable behaviour towards our only planet is bringing severe weather conditions, which will not be merciful but harsh on those underprivileged. Thank you for an inspiring book.

Iyad Abumoghli, Director of Faith for Earth, UN Environment Programme

Both enthralling and scholarly, *Weather, Religion, and Climate Change* is a timely and welcome contribution to the discourse surrounding climate change that helps us to locate weather properly in our worldview and explore the existential questions concerning weather at a time of increased vulnerability to rising temperatures and extreme weather events. In his opening remarks, Bergmann describes the book as providing us "with an exploratory trajectory, where altering weather moves from and through arts and religion to history and science, and further on to architecture and our demanding efforts to re-interpret our human place in the universe". Employing a hermeneutical lens in the phenomenological sense, he skilfully presents a fascinating and rich palette of precedents to consider across many disciplines that encourage and invite the reader to intimately experience "what it means to be bodily alive in weather lands". Bergmann challenges us to strive for "a deeper weather wising in cooperation with modern science", in order to avoid the folly of thinking we can simply geo-engineer our way out of the climate change crisis. Instead, he calls upon us to embrace a spiritually inspired reverence for weather, anchored in wonder and awe, as a critical way towards relinquishing proliferation of a dysfunctional Anthropocene era in favour of a more functional Ecocene era where humans learn to live in mutual harmony with, and respect for, all of creation.

Roberto Chiotti, Founding Principal, Larkin Architect Limited

Sigurd Bergmann's book is an insightful investigation into our consciousness of weather and the impacts and meanings of climate on human culture. The book reflects the emerging sensibility to the neglected and suppressed aspects of lived reality, and it illuminates the significance of embodiment and sensory experience, as well as the multitude of contextual, relational, and temporal processes. Modern consciousness at large has focused on forms and objects, but the evolving new awareness arises from the recognition of complex, unfocused, peripheral, and diffuse phenomena. This book also exemplifies the significance of overall views across the entire human culture, as opposed to narrow specialisations and categorisations.

Indigenous building cultures throughout history have built in harmony with local conditions, resources, climate, and weather. However, our technological culture bypasses such specific conditions and continues to develop ways of building that eliminate climatic variables through ever more complex technical systems and increasing use of energy. Besides, the abstract formal ideals of contemporary architecture have suppressed the expression of the realities of location, climate, and weather, as well as traces of ageing and use. Even today's ecological orientation in building has a technological bias, instead of seeking to collaborate sensitively with principles and conditions of nature.

Juhani Pallasmaa, Architect, Professor Emeritus, Writer (Helsinki)

This book is a major contribution to our understanding of weather through the lens of the humanities. Beautifully written and brilliantly argued, Sigurd Bergmann has created a masterpiece for us to reflect on for years to come. We are all in his debt.

Mary Evelyn Tucker and John Grim, Yale Forum on Religion and Ecology

Weather both surrounds us and saturates our being. Throughout the ages and around the world, people have lived by it, become wise to it, depicted it, revered it, and brought it into their ways of building. To assemble this weather wisdom between two covers takes a prodigious effort of discipline-spanning scholarship, never before attempted. Not only has Sigurd Bergmann succeeded magnificently in the endeavour; he has also used the resulting synthesis to take down the claims of meteorological science to have got the measure of atmospheric phenomena. Instead of the Anthropocenic alternatives of total technological control or climate apocalypse, Bergmann offers the hope of an Ecocene to come. With competing narratives of climate change, and the world on the brink of momentous transformation, this is a necessary book with lessons for us all.

Tim Ingold

Talk about the weather – wobbling between the mundane and the apocalyptic – summons in Bergmann's extraordinary meditation an atmosphere of Earth-embracing creativity. Art and climate science here collude in the spiritual force of weather to effect a "new global space", an Ecocene of surprising adaptations.

Catherine Keller, George T. Cobb Professor of Constructive Theology,
Drew Theological School. Author of Political Theology of the Earth:
Our Planetary Emergency and the Struggle for a New Public

The *Routledge Environmental Humanities* series is an original and inspiring venture recognising that today's world agricultural and water crises, ocean pollution and resource depletion, global warming from greenhouse gases, urban sprawl, overpopulation, food insecurity and environmental justice are all *crises of culture*.

The reality of understanding and finding adaptive solutions to our present and future environmental challenges has shifted the epicenter of environmental studies away from an exclusively scientific and technological framework to one that depends on the human-focused disciplines and ideas of the humanities and allied social sciences.

We thus welcome book proposals from all humanities and social sciences disciplines for an inclusive and interdisciplinary series. We favour manuscripts aimed at an international readership and written in a lively and accessible style. The readership comprises scholars and students from the humanities and social sciences and thoughtful readers concerned about the human dimensions of environmental change.

Weather, Religion and Climate Change

Sigurd Bergmann

Routledge
Taylor & Francis Group

LONDON AND NEW YORK

earthscan
from Routledge

First published 2021
by Routledge
2 Park Square, Milton Park, Abingdon, Oxon OX14 4RN

and by Routledge
52 Vanderbilt Avenue, New York, NY 10017

Routledge is an imprint of the Taylor & Francis Group, an informa business

© 2021 Sigurd Bergmann

British Library Cataloguing-in-Publication Data
A catalogue record for this book is available from the British Library

Library of Congress Cataloging-in-Publication Data
A catalog record for this book has been requested

ISBN: 978-0-367-35880-8 (hbk)
ISBN: 978-0-429-34240-0 (ebk)

Typeset in Times New Roman
by Apex CoVantage, LLC

Contents

Figures

Foreword

It is an honour to be invited to write a foreword for Sigurd Bergmann's intriguing and extraordinarily multidisciplinary book *Weather, Religion, and Climate Change*. At first when he told me that he was writing this book, I imagined something along the lines of an eco-theology of weather. But in reading it, I discover he has written something altogether more imaginative, interdisciplinary, and encyclopaedic. Ranging from landscape painting and architecture to moral theology, social psychology, economics, and climate science, this book positively shimmers with atmospheric and weather events and with aesthetic, emotional, and spiritual responses: not only on the written page but in the many fine illustrations on which Bergmann comments insightfully throughout. That weather and religion are co-implicated is both obvious in one sense and yet at the same time under-theorised. Just how they are implicated is evident in the early history of religions. The first religions, of which we have evidential knowledge, through ongoing oral traditions of those which survive to this day, were forest religions in which weather was mediated by tree canopies which filtered dappled sunlight. In forests, most indigenous ontologies perceive(d) a panoply of beings who people the environment, and they rarely give special place to a higher god or power: all beings are in some senses sacred, and the category "being" does not name a hierarchy in which gods or persons are evidently superior to jaguars or scorpions but rather an interconnected community of beings who are mutually dependent on one another. When forest humans hunt to eat, they traditionally pay reverential respect to those beings they take for their survival, and they are careful only to take those whom they know the species can bear to lose without being depleted. And when they clear forest areas for their temporary dwellings, or to grow grain or vegetables, they offer sacrifices and rituals to the spirits of the trees they remove in order that they may not turn against them. But, as recorded in a vestigial way in the biblical story of exile from the forest of the Garden of Eden, when early humans learned that they could live almost entirely from plants they selected to grow, and by modifying the habitat to grow surpluses of these plants, they cleared the forests. In Genesis 2 it is said that the occasion for the momentous exile of Adam and Eden from the Garden was plant selection: eating fruit from a tree said to be forbidden for them. Instead of a protective canopy mediating weather through a multiple ontology of beings, agrarians learned a new sense of dependence on the higher beings – the sun and the moon, rainclouds, and wind – whose activities and behaviours determined the outcome of their agricultural endeavours. And so in those first great agricultural civilisations of Babylon, Egypt, and Persia evolved the hierarchical religions of the "big gods", and, in their midst, the Hebrews discerned their own particular "god of gods".

Most religions – including the Abrahamic faiths – are peculiarly connected to the visibility of the sun, the experience of living under the sky, instead of trees, and hence to weather. And so it should not surprise us that among the early myths of the Hebrews are a slew of weather stories which shaped their sense of calling, their memories of journeying, and their settlement in the Land: the Joseph saga of the "fat and lean" years of crop surplus and famine; the Egyptian weather plagues of frogs, locusts, red rain, and so on; the great wind-driven trough in the Red Sea; the "rain of mana" in the wilderness; and the pillars of cloud and fire. Yahweh speaks, and appears, in a cloud to Moses when revealing the Ten Commandments, and Elijah, and later Christ, are said to ascend on clouds when they make their journeys back to heaven. Christ's death on the cross is accompanied by extreme weather-related events, including an earthquake and an eclipse, and the calling and missionary journeys of Saint Paul, which extend the gospel of Christ from Palestine to far-flung parts of the Roman Empire, begin with bright lights and a voice in a cloud and end with tempests on the ocean and a shipwreck.

If the Hebrews were, as Herbert Butterfield argues, the first people to read the mind of their god in the weather-shaped events of their history, they were only the first.[1] It became normal, as people lived under the sky and the sun and the stars, to read human intentions, morality, and political upheavals into the behaviours and movements of the principal heavenly bodies and the weather they generate. When Henry VIII divorced Catherine of Aragon and married Anne Boleyn, it rained for most of the following year, crops drowned in the fields, the people were hungry, and they blamed the King for putting away his holy wife. When there was an earthquake close to the Yangtze Dam in China a few months after its completion, the people said the government had lost the "mandate of heaven".

But as Bergmann so beautifully explains in this *tour de force* of a book, science has trained modern humans no longer to find meaning and morals in the weather even if it still affects their mood, nor to believe that a king's divorce, or a dictator's hubris, can bring calamity from the heavens. For 200 years of cultural transformation humans were educated, like good Newtonians, no longer to read divine or human intentions in the behaviours of heavenly objects. When the dark Satanic mills of the Industrial Revolution blotted out the skies of Manchester, Glasgow, and London with black smoke, people were told it was the price of progress, even as Ruskin fulminated against Manchester's "devil darkness" from Coniston in the Lake District, and life expectancy declined in those forced to live and die in those cities. The rich could escape the black skies to their country houses, and it was they who were the patrons of many of the artists whose work Bergmann uses in this beautiful book. But the factory workers under blackened skies got rickets, and so did their children, because there was no sunlight to give them vitamin D, and at the height of the Victorian era labourers in England died on average at just 30 years of age.

Today in industrial cities like Shenzhen and Delhi, there are again similar atmospheric conditions. But now science reveals that those conditions are toxic not only for humans but also for the Earth herself: and that the injustice that consigns the poor to die young from polluted air while the rich live in country houses enjoying the fruits of their investments is an injustice that will afflict future generations with even more intolerable weather events and that even the rich may not escape. And climate science, as well as the Gaia theory, now reveals (again) that human intentions *do* influence the weather, and that early modern scientists were wrong to imagine the Earth as a machine. The Earth is alive and is responding in increasingly dramatic and unpredictable ways to what humans have done to her since the start of the Industrial Revolution. The Hebrews, and Bergmann, are right! Weather and religion are connected by the moral law, and the moral law is not merely a cultural artefact, as philosophers from David Hume to Richard Rorty have argued. The laws of peoples and the laws of the Earth are connected. Mentality and meteorology are connected. As Werner Heisenberg first discovered, the observer and the observed are in a quantum relationship. But there are testable limits to that relationship when it comes to modern humans and meteorology. The Global Circulation Models that are run on banks of computers on four continents have turned out to be remarkably reliable at predicting the temperature increases that equate to the quantities of greenhouse gas emissions humans are intentionally injecting into the atmosphere. There is growing precision in national and company accounts about the amount of carbon dioxide that coal, gas, and oil facilities are putting, and will put, into the atmosphere. But Charles David Keeling's beautiful and portentous measure of CO_2 from the observatory in Mauna Loa, Hawaii, since 1958 keeps going inexorably up. Mechanistic science was too successful: moderns think that if science created the problem, and science can measure the problem, then science can solve the problem. The core message of this book is that science alone does not save. The scientific truths that have become "real" for moderns are no different than religious ethics and the moral law: they are knowledges that need cultural mediation, habituation, and feeling. Compassion means at root to feel *with*. Very young children in group experiments have an innate sense of justice and compassion and will seek to balance out unfairness and comfort its victims. Culture and education train humans to neglect these deep-seated moral responses. And science too has played a part in that training. All the world's religions teach the importance of compassion, and non-harm, and the duty to care for victims. But as refugees from climate-challenged regions turn up on the borders of once-Christian nations, they are being turned away. My own country Britain leads the way, despite its long history of climate pollution, in resisting the claims of refugees.

Ill will, and not technical impossibility, is the cause of the contemporary failure to respond to the increasingly extreme weather that industrialism is visiting on the Earth. The desire for justice is at risk of perishing – justice for the Earth and justice for the victims of extreme weather – and most of all in the nations that have turned away from active worship of God and gods towards the devices and desires proffered by science-informed consumerism. Weather and religion are connected. Religious ethics – an ethics of justice and compassion that is both cosmic and that arises from the heart – already points the way to moderate industrialism's increasingly overwhelming impacts on the climate systems that are driving ever greater weather extremes. This book is a treasure trove that enlightens this increasingly urgent spiritual and planetary exchange.

Professor Michael S. Northcott, Universitas Gadja Madah,
Yogykarta, Indonesia, 2 April 2020

Acknowledgements

A book like this could never have come about without a close dialogue with others. Crossing borders between disciplines, discourses, and themes is necessary in order to reflect the depth and complexity of environmental problems in general and the weather in particular, and I therefore want to thank all the scholars – in environmental humanities, our field of religion-and-environment studies, ecotheology, climatology, meteorology, and architecture – who have constructively and bountifully shared their thoughts, experiences, comments, and reviews with the author in volatile weather. Dieter Gerten, Michael Northcott, Undine Frömming, Tim Ingold, Markku Rummukainen, Tomas Björnsson, Mădălina Diaconu, Christian Hoiß, Panu Pihkala, Roberto Chiotti, Annette Homann, Helmuth Trischler, Christoph Mauch, and Upolu Lumā Vaai – my cordial thanks to you all for exciting, enlightening, and highly thought-provoking exchanges.

My stay as a fellow at the Center for Advanced Studies at Munich University in Autumn 2018 allowed me to focus intensively on the book's final chapter, and the valuable and inspiring discussions with scholars from other fields in the environmental humanities in Munich, and especially with Markus Vogt, spurred on the ideas within this text significantly.

I am deeply grateful to Marilyn Burton in Edinburgh for her skilled and dedicated work of language editing that has (once again) lifted my non-native style to unimagined heights. The Faculty of Arts at the Norwegian University of Science and Technology in Trondheim provided through its research committee a generous financial contribution, for which I am thankful.

Finally, I would like to express my gratitude to Routledge editor Rebecca Brennan, who encouragingly and professionally supported and looked after the publication process from the beginning, and to the series editors Scott Slovic, Joni Adamson, and Yuki Masami, who have dedicatedly partaken in the work's finalisation.

Note

1 Herbert Butterfield, *Christianity and History*, 98 as cited in Michael S. Northcott, *A Political Theology of Climate Change*, Grand Rapids, MI: Wm. B. Eerdmans 2013, 189.

1 Being alive in weather lands
Preliminary remarks

Mystery and ignorance

Weather belongs to the essential conditions of our bodily life. Being bodily alive is a process that takes place in the open, inhabiting a world of becoming. The world we inhabit, "far from having crystallised into fixed and final forms, is a world of becoming, of fluxes and flows".[1] In short, we live in weather lands, within a larger surrounding "weather-world".[2] Nevertheless, even though weather impacts us all, in every place and at every moment, it seems difficult to embed and locate it properly in our worldview and self-understanding. Certainly, weather forecasts are regarded as so important that they are located directly after the political news in media reports. And certainly many people listen carefully to the meteorologists' prognosis about what awaits us for tomorrow. Again and again it fascinates anew: to see and listen to someone who intends to predict the future, my future, our common future. The climate is without doubt "the theatre where human existence, humanity's history, is taking place".[3]

Furthermore, ongoing climate change and the public discourse about it turns our awareness to weather patterns. What do they look like? Are they the same as we are used to or have they changed? Increasing weather-related disasters such as droughts, floods, heat waves, deglaciation, hurricanes, and heavy rains strike populations and increase their vulnerability. Large areas of the planet's settlements and farm lands will inevitably become derelict in the near future and the waves of migration will escalate dramatically. What is good weather for whom and where? Almost everyone on Earth has learnt the lesson about dangerous climate change in the last few years, and so-called climate sceptics seem to be on the decrease, even if they can still successfully slow down or obstruct necessary transformations in social systems, technologies, and lifestyles. Many former sceptics, though, have converted to a position where they accept the climate change challenge but propagandise its solution as being exclusively through technology, without the need for any change in lifestyle. In this way, the dangerous process of sceptical denial, down-playing, and deception continues, as we self-evidently need integrated transformations in both spheres: lifestyle *and* technology.

In spite of all our technological know-how, we nevertheless still use our senses to practise old-fashioned weather wising. We look at the sky with fear and trembling, wondering how the next hours or days will shape up. Or do we stand to lose this skill by continuously looking at the screens of our mobile devices, replacing our own eyes with the weather app's poor symbols? Or can the weather app perhaps also promote such a feeling of fear and trembling?

And if "religion" is defined, as it was by Rudolf Otto in the beginnings of the discipline of religious studies, as "Mysterium tremendum et fascinosum" (a terrifying and fascinating mystery),[4] one can wonder if anthropogenic climate change is moving us towards the threshold of a new weather religion. There is at least no lack of incantations of climate apocalyptic in literature as well as among the most initiated scientists. But to whom should one pray in this regard? And for what?

Might Søren Kierkegaard be right in his *Fear and Trembling* that "infinite resignation" is the necessary condition for faith?[5] Would that mean that the insight into the anthropogenic skill to change climate confronts us with our own blindness and sinfulness? Does this make it necessary for us to give up all our earthly possessions and ultimate concerns in order to enter a state of what he calls "infinite resignation" in fear and trembling, before we, naked so to speak, are able to believe against all odds? Might changing weather, in such an existentialistic setting, offer some kind of catharsis? Is there a God who preaches through weather, or does weather simply offer us "the mirror of nature"?[6] Does weather care about humans at all?

In spite of the rise of weather-related challenges and the growing awareness of these challenges in literature and media,[7] describing weather as an underrepresented issue seems nonetheless adequate if one looks at science and the humanities. Meteorology represents weather in the scientific community, although in a narrow and instrumentalist way which we will explore later. Meteorological data and models are certainly used in other disciplines such as biology, geosciences, and physics, which treat weather empirically simply as a range of so-called data that can be classed with other data and arranged according to systems with particular theoretical interests, where the overarching epistemological question determines the answers found in nature. The complexity and ubiquity of weather seem to have quite a weak representation in the sciences. While weather does not cease to fascinate in ordinary life, science instead encounters it as mere data in a metric system.

The humanities have little more to offer. Certainly, weather appears in history as a part of different narratives about human ecology through the ages. And certainly weather was registered as an important "fact" that was able to impact on warfare and power negotiation as well as on people's potential to cultivate and survive. But recently for the first time the establishment of the academic study of climate history has imparted to weather some kind of an intrinsic value, which has made it possible to reflect on weather in its own right. Philosophy has through the ages, due to its roots in antiquity, taken place as a rather disembodied reflection on human life, although the 20th century, mainly through phenomenology, has moved the human body upwards on the agenda. Still, it seems unclear how a philosopher might interpret the significance of weather. Nor will one find much about the weather in environmental philosophy, despite its promising development in the last decades.

Literary studies naturally reflect the authors and texts which are its subject; in these, weather appears from time to time, serving a variety of functions in the narration or poem. Ecocriticism can therefore offer valuable insights into authors' rich expressions of weather contexts, and it has just recently begun to focus on weather's impact on writing.[8] The situation in art history is much more promising, even if it seems as if weather was first "invented" as a central theme in visual art in the Romantic period. Painters are, in my view, pioneers in exploring weather and its impact on nature and humanity. Imaginations and images of weather in the arts will therefore play a central role in our exploration later on in this book.

Cultural studies have also collected many observations with regard to the interaction of weather and culture.[9] Weather rites have been observed and beliefs in weather gods and spirits have been described. But here too my overall impression is that weather serves for anthropologists as a screen for projection rather than as a life-giving force in itself, even if Tim Ingold has recently started to challenge such a limited perspective.

Social sciences tend to consider the field of weather the domain of natural science, and so pay it little attention. Sometimes they produce fragments of knowledge about how adaptation to weather impacts on social (including economic) processes. In general, social science has difficulties with nature and environment even if scientists in these spheres secure considerable funding for research on sustainability, climate, green economy, etc. To put it bluntly, social scientists mostly tend to approach nature in a reductionist, inversive, and static way, and study images of nature in social beings rather than the interaction of nature and society. The currently growing discourse about the so-called new materialism might offer a promising path out of this, even if its critical and transformative potential is still somewhat unclear.[10] The well-known gap between science and the humanities is, however, still intact, and it radically hinders the study of how weather impacts on our social, cultural, and spiritual life.

My own discipline, theology and religious studies, follows the mainstream and has not much to offer either. However, here we can at least find some valuable observations, in the history of religion, in church history, and in biblical studies. Naturally, religious belief systems of all traditions imply an interpretation of how the environment and weather within it reveal God, gods, and the spirits, and how believers should respond to these. But even if the last 30 years have brought an enormous development of the study of religion and the environment, weather still does not receive much emphasis in itself. Such an exploration of weather and religion seems especially important after the recent successful establishment of the field of climate impact studies and religion. Climate science has started to dive deeper into the human and cultural dimensions of the Anthropocene; religion here offers a kind of microcosm of the complexity of human life and humans' perception, thinking, and acting. The intensification of a deeper cooperation between natural sciences, social sciences, and the environmental humanities is therefore without doubt of radical importance, and weather surrounding and breathing us all might hereby offer exceptional and unique potential for accelerating transdisciplinary multifaceted scientific reflection. As climate impact studies focus on the large-scale complex interactions of different natural processes, weather comes much closer to ordinary life and the human body and mind. The challenge to integrate weather and climate more deeply is therefore obvious for many compelling reasons. The last section of this book intends to contribute to such integration.

Weather within and all round

Even if our built environments and our primarily indoor activities mean we are little affected by weather conditions, by contrast with those working in the fields 150 years ago, weather continues to fascinate and to enchant.

The ordinary language used for talking about good or bad weather creates an illusion of a relationship between the human and the weather. Of course, there is no weather that is either good or bad; weather simply is as it is. Weather is simply weathering. It does not care about humans. It can be neither controlled nor mastered, even if geoengineering cherishes hopes of achieving such a power and awakens desires for total control over our environment. Weather simply does not take humans into account.

Just as human life is dependent on light, which surrounds us and makes it possible to see and perceive, to orient and to move and act, weather too simply surrounds and embraces us. It is "the very temperament of being".[11] According to Tim Ingold, the flux of wind and weather reminds us that we are alive in an open world:

> In this mingling, as we live and breathe, the wind, light and moisture of the sky bind with the substances of the earth in the continual forging of a way through the tangle of lifelines that comprise the land.[12]

Weather, according to such a perspective, is not just a surrounding physical element but is fundamental for every living being which breathes in air. Living in a weather world, every being is destined to combine the elements of weather in the continuation of existence.

To be alive in such a sense means to exist within the weather, to be exposed to sun that shines, to rain that falls, to wind that blows. Many humans, although protected from direct exposure to outdoor wind and weather, are still deeply affected by weather changes, even indoors. Weather conditions impact on our well-being and on our mental as well as physical sensitivities. "Being exposed to the weather and being unwell from it" is expressed in German with the appropriate adjective "wetterfühlig", to be emotionally connected to the weather.[13] To allude to Goethe, do we only know ourselves as far as we know the weather, which we only become aware of in ourselves, and is it only in the weather that we become aware of ourselves?[14] Is weather something that takes place as much within the human as around her?

Our modern understanding of weather in the lens of science has no long history. It seems to be rooted in the systematic observation of clouds in the sky, which Luke Howard pioneered in the 18th century. In a famous poem Goethe honoured Howard for his heroic feats, and re-reading it we can still sense how dramatic it must have been for our ancestors to see scientists striving to turn the uncertainty and unpredictability of weather into a rationalised system.[15]

Meteorology was certainly established by Aristotle in his work of the same title, but this only loosely collects a couple of observations without really systematising them, and without any intention of making weather predictable. Weather for Aristotle remains embedded in the movement of the stars, which he regards as divinities, and his meteorology elaborates the existence of weather within the divine configuration rather than dissecting it as modern meteorology does. Aristotle refers to his philosophical forefathers, the pre-Socratic thinkers, and he is primarily occupied with inscribing weather into the scheme of the four elements – fire, air, water, and earth – and with locating it in his overarching view of movement.[16] Weather change therefore represents for Aristotle a natural part of the world's bodily space (soma, *Weltenkörper*) and as such an outspring of the first motion, which again is anchored in the unmoveable origin of all. It may still be interesting that Aristotle was clearly emphasising life as taking place in the space between Earth and sky even if he regards this as a consequence of the divine movements in the upper world sphere. Meteorology, in its classical as well as in its modern version, is capable of maintaining the old wisdom of being alive in the fragile zone of being in between Earth and sky. In Shelley's well-known words,

> I am the daughter of Earth and Water,
> And the nursling of the Sky;
> I pass through the pores of the ocean and shores;
> I change, but I cannot die.
>
> (*The Cloud*, 1820)

Weather takes place as what Ingold circumscribes as "the temperament" of being in the flux between Earth and sky.[17] It takes place as part of the body of the world around us and vivifies the human body from within.

A barometer of God's love and morality

As weather reveals one of the most open, unpredictable, and uncontrollable dimensions of life, its uncertainty has been interpreted in the Jewish-Christian tradition as an elementary screen for interaction between creation and the Creator. As such, weather, although it certainly does not do anything other than weathering, has also served as a screen for the projection of God's presence and moral relation to his/ her created beings. In one common view, weather has been understood as the most just and equal gift of God to all on Earth, as sunshine, rain, and wind are given equally to all, and as weather does not make any distinction with regard to those that it nurtures.

> He [your Father in heaven] causes his sun to rise on the evil and the good, and sends rain on the righteous and the unrighteous.[18]

Weather is in such a view an expression of God's love to creation and his practice of sharing equally both the gifts and the challenges of life without any consideration of the individual. As everyone can be struck by (good or bad) weather, everyone is equally valued and loved by the Creator.

According to such a religious code, weather, on the one hand, represented respect towards every person. On the other hand, disasters and catastrophes represented a punishment for sin where humans did not fulfil their tasks as images of God, and where the relation between God and man/woman was broken. Injustice, lack of solidarity, oppression of the poor, and violence against one another resulted in God's pedagogically intended reaction, which was revealed by dramatic weather change. Through the uncertainty of weather, God stayed in touch with his/her created world. Weather served as a natural stage and screen for reading the Creator's relation and interaction with creation. It offered a kind of moral barometer. The relationship of morality and weather was sometimes violently intimate, such that so-called weather witchcraft and specific weather witches were blamed for catastrophes such as rain and floods, thunderstorms, and bad harvests. From classical antiquity, through the medieval period, and at least up to the early Enlightenment, women have been blamed for causing dangerous weather,[19] and the tendency is still present today in other parts of the world.[20]

While modern empiric scientific meteorology, as Ingold has shown, mainly represents the inversion of knowledge,[21] the religious interpretation keeps its eyes wide open while meeting God's eye and reading God's feelings and thoughts in the weathered book of nature. Today such a code feels strange in an enlightened world, even if extreme weather can still be experienced emotionally as something inexplicably powerful. The old moral code, however, is still intact even if we take God out of the argument. By replacing the Creator with *homo faber*, the human as engineer, the damage of extreme weather as connected to social and technical structures would then reveal the lack of perfect control. In such a technocratic culture, experts and politicians are then expected to remedy defects by establishing new machines and systems. Geoengineering, weather modification, and climate engineering can on such a view be interpreted as a historical sequel of the old code's story, where it is now the engineer who takes the place and role of the divine Creator to care for and cure creation, a creation that is now regarded and treated as a machine.

The increasing consciousness of anthropogenic climate change and our increasing vulnerability with regard to uncertain weather conditions have in some regions continued to be interpreted along the paths of old religious codes, as strikingly summarised in Michael Northcott's book title *A Moral Climate*,[22] but not many today would regard global warming as God's punishment for an unjust and unsustainable sharing of the life-sustaining resources given to us on the planet. Rather, we are looking for rational, social, and economic reasons in our own mismanagement.

Even if climate impact science again and again claims that weather is one thing and climate another, human beings as nurslings of Earth and sky and bodily beings upheld by wind and weather need to have a chance to experience the power of climate change in weather lands and contexts of dependence and empathy with the weather. Climate science that operates with large scenarios on a global scale seems to be unable to assist a transformation of insights from the global to the local, into concrete life worlds and lived spaces where weather empowers the living. Both art and religion seem to have better conditions for fertilising such an adaptation to dramatic and dangerous environmental change. Art therefore need not necessarily serve only as an illustrator of rational inverse climate science but can also follow its own traditions and foster the senses with regard to the perception of the environment, and especially its weather change. Religion need not necessarily only serve as a moral imperative that transforms normative conclusions from climate science into mobilising behaviours for establishing what scientists would regard as more sustainable. Rather, religion can mobilise its own skills to interpret the God of the Here and Now and to explore the Spirit who gives life in manifold liberating patterns. The richness of religious language emphasising weather as a spiritual force would, for example, enrich the tools for interpreting change and for creatively adapting to it in a maximally constructive way. What might weather awareness mean if it serves as a central element in religious and artistic attitudes to life and our being alive in weather lands?

What is "weather"?

Before I sketch the disposition of the book and summarise its chapters, the reader and author need to reach a consensus about what we mean by "weather". The word's content seems to be clear for most and it is used frequently in ordinary language. Nevertheless, the term has become more and more blurred for me while preparing the thoughts for this book over the past years.

In a similar way to terms like summer, night, nature, and religion, the notion of weather also represents an abstraction that agglomerates different elements. The term persuades us that there is something like weather, but while thinking and talking about it, we must scale down to phenomena and words such as wind, rain, sunshine, and humidity. If there "is" something like weather, it always appears as a synthesis of several different states of being. "Weather" is not only a natural phenomenon; "weather is a rhetoric", as Lisa Robertson aptly reminds us.[23]

The word has its roots in the old German word "wetar", which means wind and blowing, breezy weather, flying. In German, "weht der Wind" (the wind blows). Weather, we might say, is originally what we feel as a movement of air on our skin, hot or cold and wet or dry. Weather is what affects us bodily in our environment.

Scientifically, weather is defined as "the condition of the atmosphere at any particular time and place".[24] The atmosphere is for the scientist an enormous thermodynamic system that surrounds the more stable Earth.[25] What science defines as weather is simply the state of the atmosphere, with regard to different parameters; and as the weather is highly changeable in time and place, meteorology studies the variability of these conditions. Two elements are, as we can see, central for the scientific understanding of weather: the spatiality of the earth's atmosphere as a cosmic place and its changeability in time and place. Later we will explore more deeply how change and spatiality are understood and represented in the language of meteorology. To begin with, it is enough to detect the sameness and difference between ordinary and scientific language. While the scientist observes empirically a range of weather elements in the atmosphere, weather in ordinary language aims at the bodily experience of one's physical surroundings and the experience of variability. The usage of the term "weather" in ordinary language seems to be in unexpected, but also deceptive, conformity with its scientific use, even if it can be loaded with a normative meaning where weather can be good or bad (for us), which is nonsense in a meteorological sense. In scientific as well as in other life worlds, the central significance of weather is its changeability. Again and again, one has to look at the sky to identify what is happening and to remember and compare with how it was before. One might say that weather in this regard demands of humans a continuously ongoing awareness of it and skill to empathise with it. Furthermore, weather makes incredible demands on our memory and ability to remember. Weather embeds us in our surroundings, and it is the steady interconnection of our body and mind to our environment. Weather takes place around and within us. It is there all the time and in every place.

While the scientist is interested in the study of only a range of selected weather elements, the poet can treat weather as a blank canvas. Poetry can "paint" life on the weather screen and load it with manifold rich meanings. Rainer Maria Rilke, for example, often takes colours as a starting point in expressing nature- and weather-based moods. In "Before Summer Rain", he depicts in masterly fashion the mystery of weather change:

> *Suddenly, from all the green around you,*
> *something – you don't know what – has disappeared;*
> *you feel it creeping closer to the window.*[26]

As rain comes closer, we become aware that something disappears, and the more it creeps towards us, the more difficult it becomes to remember and "know what has disappeared". Science, due to its empirically limited methods, can only mirror and measure what it calls "change". But the poet is able to explore the depth and metamorphosis of nature and human life *in between* the within and around. Change in Rilke's poem is not simply the movement from one (static) state of being to another but it is the back and forth, the flux of weather, the temperament of being. Its movements are neither linear nor regularly circular but rather rhythmic like a wave. In addition to this, the movements of weather are layered.

As rain creeps closer, sunlight disappears. As our awareness concentrates on the what-is-not-yet-there, our memory undergoes a shift in reinterpreting the past in the light of what is emerging. While adapting to new weather, we rewrite our past. From the poet we can learn that weather shifts are never simply displacements from one frozen state to another but take place as an ongoing complex polyvalent movement which rather follows musical principles of harmony, consonance, and dissonance. Weather can sound, metaphorically as well as physically. Weather also paints in colours. The grey sky turns to red. The dark blue transforms into the bright white. Wind blows whistling in the leaves and rain patters on the roof. Weather appears as humidity on our skin but it also sounds in our ears and manifests in colours in our eyes. It continuously captures our awareness, perception, thinking, and acting. Therefore, only a synaesthetic[27] approach to weather, and its religious imagination, seems reasonable.

Disposition

I hope that the reader can follow me so far and abandon the taken-for-grantedness of weather. Even if meteorology and ordinary language pretend a clear, obvious, and unambiguous meaning of the term, it offers a rich and still little explored source for reflection on how the human body and mind interact with the surrounding nature and its flux.

In the following, I will proceed in eight chapters, which are ordered in partly overlapping thematic sections of art, culture, religion, history, climate, Anthropocene, Ecocene, and atmosphere. As visual artists have approached weather's mystery of alteration and its interplay with the human body in a different way from scientists, we will start by seeking inspiration in selected paintings from one of Europe's masters who was painting at a time when our modern perception of weather in land and seascapes was taking shape. It is not without good reason that J.M.W. Turner has been called Master of Light, and here he also appears as Master of Weather. With senses open to the mysteries of weather alteration, we will then turn from art to faith and religion, where excursions will take us to a broad range of different religious belief systems in various times and regions of the world, before we explore more deeply how Jewish-Christian faith encountered and interpreted serious weather changes in classical, late medieval, and early Reformation history. The historical exploration of religion is then broadened to the history of weather science from antiquity to the Enlightenment, when modern scientific meteorology takes its course. In order not to let science, and our critical and constructive evaluation of modern meteorology, have the final word, the place and built environment where weather actually happens comes into focus in a reflection on shelter, faith, and architecture, where several lines from earlier chapters flow together. Insights from all chapters are assembled in the final chapter, which explores their relevance and significance in the ongoing discourse about the Anthropocene, where atmospheric thinking in a meteorological, philosophical, and spiritual sense allows us to imagine not the end of this world but a time and Earth beyond: the Ecocene. What at first glance might have looked like quite a random structure may now appear as an explorative trajectory, where altering weather moves from and through arts and religion to history and science, and further on to architecture and our demanding efforts to reinterpret our human place in the universe. The golden thread of the booklets weather and its power and mystery of alteration blow through all these spheres of reality and encourages the reader to experience what being at its mercy might imply.

In the following, I present the chapters' content in more detail. As artists have approached weather flux in a complex way (though differently from scientists), I will begin in Chapter 2 by seeking inspiration in selected paintings of J.M.W. Turner, who has rightly been honoured as the inventor of weather in modern art. Rilke will also assist us here in finding words for artistic weather wisdom. What was it that Turner virtually explored and "discovered"? And what can we learn from his paintings with regard to the phenomena of change, flux, and variation in our own environment?

Chapter 3 collects observations in cultural and religious studies about what we might call weather images and responses. While atmosphere, as we have seen, represents the central term in meteorology, atmosphere in this chapter will be used as a hermeneutical lens in the phenomenological sense. Culture, and religion within it, responds to weather as atmospheres which are taking place "in between" – between culture and nature, between humans and gods, and between spirits and natural elements. Weather, one might say, produces atmospheres of belief in culture. A surprisingly rich variety of different symbolic imaginations of weather phenomena appears on the map of different cultural and religious traditions, and one can only wonder why religious studies so far has not put stronger emphasis on mapping these. The chapter does not claim to offer such a complete map but aims rather at encouraging others to target it. It evolves as an expedition to different locations, times, and histories: from Fiji to Greek antiquity, from "healing twins" on different continents and in different traditions to birds among Native Americans, and from the weather god Yahweh in biblical times to weather wisdom and modification in ancient Central Asia.

"Atmospheres agog" are in this chapter explored as variations of weather in continuous cultural religious movement. Religion, serving as a kind of atmospheric inversion, intervenes in this flux and allows human beings to both identify themselves at home in bodily partaking in the weather on the one hand and spiritually establishing a relation to its hidden forces on the other. Religious imagination of and response to atmospheres agog makes it possible to empower one's life through weather and at the same time strive to become an active part of its life-empowering rhythm and variation.

While later chapters focus explicitly on Christian perspectives in a Western tradition, this chapter collects observations from a wide variety of different continents and religious traditions. African, Asian, American, Pacific, and European understandings sit side by side, and local native belief systems appear together with popular Islam, classical mythology, and Jewish and Christian faith traditions.

Chapters 4 and 5 offer insights from history. Aristotle's text on meteorology was the first attempt to construct a system for reflecting on weather. In the context of this book, it is of special interest to investigate how the philosopher's understanding of the gods and stars and the understanding of movement impact on the depiction of weather.

The biblical books, both the Hebrew Bible and the New Testament, offer rich sources for the observation of weather in Jewish and Christian faith. Might it be possible that the whole image of God in ancient Israel emerged from belief in a Canaanite weather deity? How are the Scriptures expressing God's care and love to creation on the canvas of weather? How does the moral code work where weather disasters are interpreted as signs and punishments for sin on the one hand and for the education of believers on the other? And do these codes still work today even if one has no need for the *God hypothesis* in a modern secular context?

Two exciting examples of how Christian faith communities have used weather for the interpretation and explanation of natural and sociopolitical challenges are found in Christian Europe in the late medieval and post-Reformation eras. In seeking for explanations of problems in weather-dependent agriculture, where disasters were striking vulnerable populations, women were accused of being witches and held responsible for the production of "Unwetter", non-weather – that is, extreme weathers and disasters. Images and stories about so-called weather witches help us to visualise how deeply weather was understood and treated religiously at that time, but they also reveal how sexism and gender injustice offered a break with the older Jewish-Christian tradition where one first investigated what the Creator intended to preach through demanding weather rather than immediately seeking, finding, and punishing scapegoats.

Another interesting theological approach to weather is found in German Protestant sermons from the 18th century, where thunderstorms, droughts, and bad harvests are interpreted as signs from God, which are calling for repentance. Weather thus received a deeply ethical function where self-critical reflection on the part of the population was encouraged by the theological interpretation. Such a code still seems to have a central function in environmental movements where the so-called ecological crisis, and nowadays accelerating climate change, is understood as an alarm clock for modern society with a clear call for conversion, culturally, socially, politically, and economically. The genre of "Wetterpredigten and Donnerpredigten" ("weather and thunder sermons") appears here surprisingly as a kind of forerunner of environmentalism's revivalism.

The history of meteorology will be traced in Chapter 5, where the long influence of Aristotle's *Meteorologica* is particularly emphasised. The classical philosopher's central influence lasted until the invention of measuring tools such as thermometers and barometers made possible the transition from the study of weather in the expanse beneath the stars to empirical observation from Galileo and Descartes onwards. In the 19th century, one can for the first time perceive a radical de-sacralisation of weather due to scientific and technological advances that dramatically altered our perception and understanding of weather phenomena. Mythology and religion were replaced by science and rational explanation. Humboldt, Howard, and others were acknowledged to have reduced uncertainty and, in Goethe's words, to have given "precision to the imprecise", and the Romantic thinkers were striving to explore the intertwinement of the human self and the surrounding atmosphere of weather. Humboldt in particular combines a fascinating skill in measuring and empirically investigating nature, including the weather, with exploration of weather within a wide range of natural phenomena, for example in his plant sociology where topographic, meteorological, and edaphic (soil influenced) conditions are integrated.

As the scientific study of the atmosphere dominates the modern worldview, its reductionism can be criticised. In dialogue with Tim Ingold and Bron Szerszynski, the chapter investigates how modern meteorology also includes the fatal danger of perceiving weather and climate simply as measurable, mathematically readable, and techno-writeable phenomena that are under human control and that can be mastered by economic and engineering methods. Obviously, they are not. The challenge to strive for a deeper weather wising in cooperation with modern science is urgent in the present new horizon.

Chapter 6 mines this ambiguous situation more deeply, where reason has replaced belief and where weather has lost its enchantment. Meteorology can certainly analyse and describe what happens, but it cannot offer any assistance to cope with the existential dimension of weather's impact. This chapter explores what I call the commodification of weather, as it has been accelerated by the development of both modern meteorology and financial capitalism, where economic interests increasingly dominate the perception and study of weather as a traded and fetishised object. Might religions be able to mobilise a countervailing power to the commodification of life, where faith perceives weather as a sacred gift rather than a commodity? Might they be able to catalyse a view of weather as a global common and of meteorology as for the benefit of all?

From the beginning, human architecture has been closely linked to weather. Furthermore, built environments have always served as symbolic spaces for encountering the Sacred. The entwinement of architecture, weather, and religion therefore offers a highly complex, rich, and inspiring source for deepening the book's reflections. Chapter 7 explores this interconnection in a broad geographical and extended historical range. Phenomena such as weather roofs in Stone Age and vernacular architecture, medieval sacred architecture's adaptation to changing weather, and late modern eco-architecture bring us face to face with a challenging choice: to build either *against* or *with* weather, either protecting *from* weather or designing *with* it. Modern, environmentally conscious church architecture offers here an intriguing creative arena for experimenting with new modes of building with weather. The chapter leads us to the question of if and how "built faith" might contribute a driving force towards sustainable architecture in weather lands of increasingly demanding alterations. Not only buildings but the whole of urban space and its planning need to be radically revised in light of anthropogenic weather change. Historical responses to similar weather changes in the past, even if these took place in local contexts on a smaller scale, can offer unexpected exciting inspiration and lessons for how we might design sustainable spaces for our common future in a postmetropolitan Anthropocene.

Chapter 8 takes the exploration of weather in the foregoing chapters into the discourse about climate change and the Anthropocene. In harmony with the book's intention to keep its reflections close to the awareness and experience of being bodily alive in weather lands, the chapter starts with remembering the long-term heat wave that characterised Central and Northern Europe's extraordinary summer in 2018, a summer that was still going on and only slowly transforming into meteorological autumn when the first version of this chapter was written in the Bavarian Alpine foreland in late October. Envisioning the extreme ambiguous bodily experiences, feelings, thoughts, and environmental and meteorologic explanations from this summer, the chapter highlights the role and function of religion in the Anthropocene.

Taking seriously Bron Szerszynski's plea for "opening the climate" beyond the techno-writeability of meteorological weather, discussed in Chapter 5, I approach the discourse about the Anthropocene from a criticism of earth system analysis' poor concepts of change and of the human. Climatology's knowledge production, though without doubt highly successful and significant across many disciplines, is nevertheless in need of being expanded and advanced by arts, culture, and religion in order to foster a deeper understanding of weather alteration and human unpredictability and complexity, as the foregoing chapters have elaborated through following the artists, scholars, and believers we have met. Critical light will be shone on the pros and cons of the still unclear and ambiguous depoliticising Anthropocene discourse. Furthermore, Christian theology's function with regard to the discourse will be discussed, where especially the concept's lack of imagining the future is critically highlighted and constructively overcome in the horizon of a still yet unseen *Ecocene*. Following our extensive volume on "Religion in the Anthropocene",[28] faith and religious belief necessarily take place today *in* the Anthropocene. Christian liberation theology thus needs to reinvent itself as a critical creation theology *within*, and also, as I will envision, *beyond* the Anthropocene. The chapter will explore theological skills for widening our vision from the past and present to a future beyond the Anthropocene. In this way, it will explore how a move of contextual theopolitics might contribute to experiencing the earth as Ecocene.

Finally, I compress the book's reflections about the concept of atmosphere. The term, as we have seen, both has served as a central analytic term in science and meteorology and offers concurrently, in environmental aesthetics, a crucial key for overcoming fatal dichotomies in Western thinking about nature that makes it possible to think *with* nature. Can meteorological and affective atmosphere enter into a fruitful dialogue and encounter? Can the notion also assist thinking with weather in reflecting our bodily being-alive in the flux of lands of alterations? How might one stage an exchange and interplay between the atmosphere of scientists and that of the environmental humanities and arts? And what about atmosphere as an intriguing spiritual term in the world of faith and weather wisdom?

Additionally, with regard to time and human imagining of the future, a deepened understanding of atmospheres might gain a central constructive significance in the challenging context of sociocultural creative adaptation to so-far-unexperienced weather conditions. Due to the crucial power of language over our experience and interpretation of being alive in weather,[29] narratives about climate change play a central role. Narratives about how changing climate impacts on weather wield specific power over how we are negotiating our practices and worldviews. Nevertheless, in addition to narratives, images – our power of imagination as well as our power of image displaying – are likewise crucially decisive,[30] and they can determine how we respond to what we see, imagine, and envision in demanding weather alterations.

What kind of atmosphere appears in both narratives and iconographies about weather and climatic change in the public sphere, and how might these connect to religious "eschatology as imagining the end"? Does "imagining the end" aim at the final termination of Earth as we know it, or is it understood in a key that allows substantial power for the New to arise from it?

In a nutshell, what can weather teach about our common future and atmosphere on Earth, our home? Can weather wisdom nurture hope and establish practices to manifest this hope for the Ecocene? How do we cope with a more and more politicised weather (where its impacts produce new injustices and challenges, and sharpen existing ones)?[31] Can we manage not simply to talk about weather but "to talk weather"?[32] How do we make ourselves at home with weather, and how do believers encounter the Holy Spirit as Giver of Life and Weather? It once dawned on Marcel Proust that "A change in the weather is sufficient to recreate the world and ourselves",[33] and the author would be satisfied if the reader of this work might increase his/her disposition to try to let this come true through a changing awareness of weather and of being alive in it. Might weather wisdom then re-create the world and ourselves?

Notes

1 Tim Ingold, "Lines and the Weather," The Daphne Mayo Lecture, presented at the University of Queensland Art Museum on Wednesday 16 October 2013 (pre-print manuscript p. 11), https://soundcloud.com/uqartmuseum/tim-ingold-lines-and-the-weather-daphne-mayo-lecture-2013, accessed 12 August 2019.
2 Coining the term "weather lands" is inspired by Tim Ingold's coinage of the term "weather-world". According to Ingold, fundamental to life "is the process of respiration, by which organisms continually disrupt any boundary between earth and sky, binding substance and medium together in forging their own growth and movement. Thus to inhabit the open is not to be stranded on the outer surface of the earth but to be caught up in the transformations of the weather-world". Tim Ingold, "Earth, Sky, Wind, and Weather," *The Journal of the Royal Anthropological Institute* 13, 2007 (*Wind, Life, Health: Anthropological and Historical Perspectives*), S19–S38, S19. Cf. Tim Ingold, "Footprints through the Weather-World: Walking, Breathing, Knowing," in: T. H. J. Marchand (ed.), *Making Knowledge: Explorations of the Indissoluble Relation Between Mind, Body and Environment*, Oxford: Wiley-Blackwell and London: Royal Anthropological Institute 2010, 115–132.
   Living in weather lands then, in the sense in which I use it, refers to concrete local life as part of generally being alive in the weather world in Ingold's sense.

3 Wilhelm Lauer, *Klimawandel und Menschheitsgeschichte auf dem mexikanischen Hochland*, Abhandlungen der Mathematisch-Naturwissenschaftlichen Klasse Klasse, Jahrgang 1981, Nr. 2, *Akademie der Wissenschaft und der Literatur*, Mainz, Wiesbaden: F. Steiner 1981, 5: "Das Klima ist für die Gestaltung des Schauplatzes, auf dem sich das menschliche Dasein – die Menschheitsgeschichte – abspielt, tatsächlich von Bedeutung, denn es steckt im weitesten Sinne den Rahmen ab, beschränkt Möglichkeiten, setzt Grenzen für das, was auf der Erde geschehen kann, allerdings nicht, was geschieht oder geschehen wird".

4 Rudolf Otto, *Das Heilige*, München: C. H. Beck. 1 Auflage 1917.

5 Søren Kierkegaard, *Frygt og Bæven*, Copenhagen: C. A. Reitzel 1843. Engl. transl., *Fear and Trembling*, Harmondsworth: Penguin 1985, 70, 75.

6 Cf. Richard Rorty, *Philosophy and the Mirror of Nature*, Princeton, NJ: Princeton University Press 1979.

7 Several important media have established sections for weather information which go beyond meteorology, and national meteorological institutions have widened their services and they also include nowadays educational and popular science background information: cf. ZDF, BBC, SVT, Süddeutsche Zeitung, Dagens Nyheter, Times, Guardian, SMHI, Yr.no, DWD.

8 An outstanding attempt to focus explicitly on weather in literature and arts took place, for example, at the conference "What's the Weather Like in Anglophone Literature and Arts," 14–15 October 2016, Université Sorbonne Nouvelle, Paris, cf. https://victorianpersistence.wordpress.com/2016/10/08/conference-whats-the-weather-like-in-anglophone-literature-and-arts-14–15-october-universite-sorbonne-nouvelle-paris/, accessed 12 August 2019. Highly valuable reflections can be found in several studies such as Alexandra Harris' rich (English) history of writers' and painters' reflections on the wind, rain, and sun in *Weatherland: Writers & Artists Under English Skies*, London: Thames & Hudson 2015. A new genre related to the theme of this book is found in climate change fiction, abbreviated as "cli-fi", even if the climate (discourse) overshadows the weather. Cf. Adeline Johns-Putra, "Climate Change in Literature and Literary Studies: From Cli-fi, Climate Change Theater and Ecopoetry to Ecocriticism and Climate Change Criticism," *WIREs Clim Change* 2016, doi: 10.1002/wcc.385, https://internt.ht.lu.se/doc/1458129274.calendarEvents.8222.pdf.0.Johns-Putra_20Clifi.pdf/Johns-Putra%20Clifi.pdf, accessed 12 August 2019.

9 See, for example, Vladimir Jankovic, *Reading the Skies: A Cultural History of English Weather, 1650–1820*, Chicago: The University of Chicago Press 2001; Mike Hulme's, *Excellent Weathered: Cultures of Climate*, London: Sage 2016 offers another significant and influential contribution, where it is nonetheless not weather in its own right but climate (the manifold dimensions of the climate change discourse and the cultural construction of "climate") that is at the core.

10 With regard to the connection between new materialism thinking and the study of religion and the environment, our conference in Manchester in 2021 will probably offer new valuable insights: *The European Forum for the Study of Religion and Environment* in association with the Lincoln Theological Institute's *Sixth International Conference Religion, Materialism and Ecology*, 14 May to 15 May 2021 at the University of Manchester, UK. Cf. http://lincolntheologicalinstitute.com/efsre-vi/, accessed 18 September 2020.

11 Tim Ingold, *Being Alive: Essays on Movement, Knowledge and Description*, London and New York: Routledge 2011, 130.

12 Ibid., 115.

13 Interestingly, German offers a technical term in the adjective "wetterfühlig" while English lacks such a term and circumscribes the state of being. The phrase "being

under the weather" certainly draws on weather as a metaphor but aims at being unwell in general. Are the British, surrounded by rough sea, tougher in their bodily weather response than Germans? Or might one trace an interesting difference between German and English romantic language production?

14 Johann Wolfgang Goethe, "Bedeutende Fördernis durch ein einziges geistreiches Wort," in: *Zur Naturwissenschaft überhaupt, besonders zur Morphologie: Erfahrung, Betrachtung, Folgerung, durch Lebensereignisse verbunden*, Munich: Hanser 1989 (1817–1824), 306–309: "The human being only knows herself as far as she knows the world, which she only becomes aware of in herself, and only in the world she becomes aware of herself".

15 Cf. Chapter 5.

16 Aristotle, *Meteorology*, Book 1, chapter 2–3.

17 Ingold, *Being Alive*, op. cit., 130.

18 Matthew 5:45, NIV.

19 Cf. also the bad impacts of weather shocks on violence against women in Africa and their economic dependence. Maty Konte and Nyasha Tirivayi (eds.), *Women and Sustainable Human Development: Empowering Women in Africa*, Cham: Springer Nature and Palgrave Macmillan 2020, 42–43.

20 Unfortunately, I cannot provide clear evidence, but after field trips and encounters in Korea and Kyrgyzstan, there seems to be an interesting gender difference where men who are initiated in weather wising are regarded as highly respected authorities, while women are instead held responsible for "bad" weather, even though it is also possible to meet respected female weather wisers. My photo gives a sense of the sacred landscape and holy place in Kyrgysz popular Islam, where weather shamanism in connection to the holy lake also plays a prominent role: https://seeingtheforestdotorg.files.wordpress.com/2013/05/kyrgsz-rcc.jpg, accessed 13 August 2019.

21 For Ingold, "the logic of inversion" turns "the pathways along which life is lived into boundaries within which it is contained. Life, according to this logic, is reduced to an internal property of things that *occupy* the world but do not properly *inhabit* it. A world that is occupied, I argue, is furnished with already existing things. But one that is inhabited is woven from the strands of their continual coming-into-being". Tim Ingold, "The Wedge and the Knot: Hammering and Stitching the Face of Nature," in: Peter Scott, Sigurd Bergmann, Heinrich Bedford-Strohm and Maria Jansdotter Samuelsson (eds.), *Nature, Space and the Sacred: Transdisciplinary Perspectives*, London: Routledge 2009, 147–161, 147.

22 Michael S. Northcott, *A Moral Climate: The Ethics of Global Warming*, London: Darton Longman & Todd 2007.

23 Lisa Robertson, *The Weather: A Report on Sincerity*, Washington, DC: Poetry Anthology 2001, www.dcpoetry.com/anthology/242, accessed 17 October 2018.

24 C. Donald Ahrens, *Meteorology Today: An Introduction to Weather, Climate, and the Environment*, 6th edition, Pacific Grove: Brooks and Cole 2000, 15; and Helmut Kraus, *Die Atmosphäre der Erde*, 4th edition, Berlin, Heidelberg and New York: Springer 2004, 11.

25 Kraus, op. cit., 3.

26 Rainer Maria Rilke, "Vor dem Sommerregen," in: *Neue Gedichte, Erster Teil (1907)*, *Sämtliche Werke*, Vol. 1, Frankfurt am Main: Insel Verlag 1955, 520.

    Vor dem Sommerregen

    Auf einmal ist aus allem Grün im Park
    man weiß nicht was, ein Etwas fortgenommen;
    man fühlt ihn näher an die Fenster kommen

und schweigsam sein. Inständig nur und stark
ertönt aus dem Gehölz der Regenpfeifer,
man denkt an einen Hieronymus:
so sehr steigt irgend Einsamkeit und Eifer
aus dieser einen Stimme, die der Guß
erhören wird. Des Saales Wände sind
mit ihren Bildern von uns fortgetreten,
als dürften sie nicht hören was wir sagen.
Es spiegeln die verblichenen Tapeten
das ungewisse Licht von Nachmittagen,
in denen man sich fürchtete als Kind.

Before Summer Rain

All at once from the green of the park,
one can't quite say, something is taken away;
one feels it coming closer to the windows
and being silent. Out of a grove,
persistent and strong, sounds a plover,
one thinks of a Saint Jerome:
so intensely rises a solitude and fervour
out of this one voice that the downpour
shall listen. The walls of the great hall
with their paintings retreat from us
as if not allowed to hear what we say.
Reflected in the faded tapestries
is the uncertain light of afternoons
in which one as a child was so afraid.
(transl. Cliff Crego)

http://picture-poems.com/rilke/features/lochbergbach.html, accessed 23 October 2018.
27 *Synaesthetics* derives from "synaesthesia" (Greek "together" and *aisthesis*, "sensa-tion" or "perception"). In arts and aesthetics, it means that the senses cannot be sepa-rated from each other but are interacting. A colour can, for example, be experienced as sound and a number as a spatial position. Sounds can evoke colours. It is also a medical term to define a neurological condition where a fusing of sensations occurs when one sense is stimulated which automatically and simultaneously causes a stimulation in another of the senses.
28 Celia E. Deane-Drummond, Sigurd Bergmann, and Markus Vogt (eds.), *Religion in the Anthropocene*, Eugene, OR: Wipf & Stock and Cascade 2017.
29 Cf. Lisa Robertson, *The Weather*, Vancouver: New Star Books 2017.
30 Cf. Birgit Schneider, *Klimabilder: Eine Genealogie globaler Bildpolitiken von Klima und Klimawandel*, Berlin: Matthes & Seitz 2018.
31 Cf. Stefan Schmitt's poignant remark that "it never has been so political as now to talk about weather" (Nie war es so politisch wie jetzt, übers Wetter zu reden): "Politische Energie: Die Hitze drückt, aber sie könnte den Ideenwettbewerb der Parteien für den Klimaschutz beflügeln," *DIE ZEIT* 74, 27, 2019, 1.
32 In his wonderful unique novel (shaped as a narration where the author, in an interview with a literary critic, remembers writing his novel), Wolf Haas elaborates the difference between talking about the weather and weather-talking, when his lead character – for whom not a single human being is as interesting as the weather (p. 49) – meets a woman with whom he can really "talk weather" (Wetter reden) (p. 57) – that is, talk

not just about weather's impact and forecasts but also about "the final things", such as the change of cloud pictures leading to a volcanic eruption and the movement of the cap between the tropo- and the stratosphere. Wolf Haas, *Das Wetter vor 15 Jahren*, 8th edition, München: dtv 2017.

33 Marcel Proust, *Le Côté de Guermantes*, Paris: Gallimard 1921. Engl. ed., "The Guermantes Way," in: *In Search of Lost Time*, 6 Volumes, Vol. 3, London: Vintage Classics 1996, 1014; Cf. Adam Watt, *The Cambridge Introduction to Marcel Proust*, Cambridge: Cambridge University Press 2011, 67.

2 Inventing weather

Conveying the mysteries of alteration in J.M.W. Turner's painting

One of my most fascinating teachers about the intrinsic value of weather is Joseph Mallord William Turner. His paintings reveal unstintingly the inner force of nature as it appears in the atmospheres of weather landscapes. Turner's mode of painting seascapes, landscapes, built and natural environments, and impressions of flowing light visualises both the weather and our seeing and feeling of it. It allows us to become aware of the weather's physical manifestation at the same time as it makes us conscious of our perceiving of it. Turner's painted weather takes place outside, somewhere else, as well as deep within us. What happens on the canvas connects the outside with the inside. His weather lands dissolve untiringly the borders of within and without.

I would never dare to think of writing about such a national hero of British and European art history as Turner and to try to cover the complexity and scope of his work and historical significance. Too many others, who are more skilled, have already done this.[1] My intention here is simply to do justice to Turner's achievement of having given the atmospheres of weather lands such intense expression that weather since then has been able to be perceived and understood as a force that impacts on the earth and the bodily alive human being. Turner has indeed invented weather in its modern sense. In spite of the limitations of visual art, where one only can freeze a state of being in time, Turner has ingeniously been able to grasp and display the weather's changeability, or what I later will describe as its alteration. For him, changing weather offered the artist "dispensing incidents" and represented a central challenge to the landscape painter.[2] As weather does not simply change from one state to another but reveals the power of variation in itself, it demands specific artistic skills to express its metamorphic dynamics in a picture. Turner's work has obviously developed these skills in a masterful way, such that the power of natural change is unfolded on the canvas as well as in the eye and soul of the viewer. His land- and seascapes are always appearing atmospherically;[3] that is, they actively take place in between object and subject, as a synergy between nature, canvas, and eye.[4]

J.M.W. Turner (1775–1851) developed his artistic skills early in life, taking an interest in architecture but also painting water colours, a genre that he developed throughout his lifetime in a highly creative and masterful way; he cultivated, for example, a wet-on-wet-technique in a way that made it possible to experiment with the flow of light and colours in an entirely novel fashion. In the tradition of landscape painting, Turner departed from the giants of his time – especially Claude Lorrain and Nicolas Poussin, who were highly appreciated in Britain in his day – but cultivated more and more his own style where the transparency of light became increasingly significant and where almost pure light, shining through shimmering colour, fluency, and atmospherical impression, moved into the centre. Through his rivalry with John Constable, Turner's pictorial universe evolved dynamically. Seascapes and landscapes, wrecks, and architectural environments, but also historical, biblical, and mythological sceneries, as well as disasters,[5] mainly storms but also fires, quickened his interest. In his development, Turner moves from an interest in topography towards more historical and epic scenes.[6] Compared with Constable and others, Turner's emphasis was much more on the colour and light of the landscape than on its spatial depth.[7]

Oskar Bätschmann rightly notes that Turner, especially in his famous *Rain, Steam and Speed* (Fig. 2.1), dissolves the conventions of how space and perspective are portrayed by eliminating the vanishing point and thereby "destroying space".[8]

Figure 2.1 Joseph Mallord William Turner (1775–1851), *Rain, Steam and Speed, The Great Western Railway*, before 1844, oil on canvas, 910 × 1218 mm, National Gallery, London.

https://commons.wikimedia.org/wiki/File:Joseph_Mallord_William_Turner_-_Rain,_ Steam_and_Speed_The_Great_Western_Railway_-_WGA23181.jpg, accessed 17 December 2019

But I think Bätschmann goes too far in claiming that Turner hereby prioritises time over space and replaces spatial expansion with temporal.[9] Rather, it seems to be the subjective experience of speed in a weather environment that lies at the heart of *Rain, Steam and Speed*. Turner himself talked about the emotive experience he had while leaning out of the window for several minutes.[10] Rather than replacing space with time, I would talk about the artist's bridging space and time through a subjective exploration of the environment impacting a body in motion. Through this, the painting succeeds in depicting a place with its own specific character, a place of encounter between the landscape, body, and machine. Motion and the flow of time is of course a crucial element in this, but for me it seems to be one of the presuppositions for a new experience of nature within and without, rather than the emphasis of the picture itself.

From his youth Turner had a strong interest in the immaterial carriers of colour, such as dusk, rain, and fog. In his later years, the artist seems to have become even more occupied with the intrinsic value of colour and light, which is clearly impressed in his so-called colour paintings, especially after his travels in Italy, France, Germany, Belgium, Austria, and Switzerland from 1819; these works, which were produced from 1818 onwards and which Turner called "beginnings",[11] took painting nearly to an abstract and nonfigurative depiction of pure light. In an untiring way, Turner had underpinned these paintings with many independent sketches of clouds, waves, sunsets, and storms, "not topographically located but probably inspired by stays at Margate and wanderings along the Channel coast".[12] However, as the most experimental paintings were never intended for exhibition, one should, on the one hand, not overestimate his production as preparing the way for the later impressionist movement. On the other hand, Turner's experiments undoubtedly represent in themselves a break with naturalistic conventions which allows the painter to investigate the inner quality of the phenomenology of nature (Fig. 2.2).

The condescending reaction of his colleagues in the Royal Academy to his paintings which explicitly drew on his experimental colour beginnings confirms even more how Turner was more than a little before his time: "Soapsuds and whitewash" was what some critics called his famous "Snowstorm" from 1842 (Fig. 2.3). And the general evaluation of his colours was a humiliating characterisation of his pictures as "of nothing and very like".[13]

Similarly to *Rain, Steam and Speed*, *Snowstorm* is also anchored in a deep individual experience. Turner himself reports how he asked the sailors to be lashed to the mast of the steam-boat *Ariel* lying off Harwich in a snowstorm for four hours, in order "to observe it" and "to record it";[14] and he explicitly completes the painting's title with details about it. In the catalogue text, he describes himself as the painting's "author". One can scarcely come closer to being exposed to the force of weather. Due to his impressive memory skills, Turner could afterwards express and paint what he saw and felt; on the canvas he shared the feeling of being totally dependent on the awful and wild weather. One is tempted to compare his experience with Friedrich Schleiermacher's definition of religion as "the feeling of absolute dependence" and ask whether Turner continuously explores in his

Figure 2.2 Joseph Mallord William Turner (1775–1851), *Tancarville: Colour Beginning*,
      1839, watercolour on paper, 307 × 488 mm, Tate Britain, London.

Photo © Tate

Figure 2.3 Joseph Mallord William Turner (1775–1851), *Snowstorm – Steam-Boat Off a
      Harbour's Mouth Making Signals in Shallow Water, and Going by the Lead*, ca.
      1842, oil on canvas, 910 × 1220 mm, Tate Britain, London.

Photo © Tate

paintings what it means to be alive and totally dependent on the forces of nature manifested in the atmospheres of weather.[15] Eberhard Roters strikingly states that Turner in this painting locates the acting God, who earlier has intervened in nature from without, now directly in the power of the elements.

> The divine omnipotence is embodied in weather itself; the meteorological has stepped in in place of the anecdotal narration about God's action upon human fate, wherein nature's breathing becomes comprehensible as the mantle of an incomprehensible and unpredictable movement of destiny to which we are helplessly exposed.[16]

In any case, one might ask if Turner here also makes us aware of the demon spirits of technology who are moving through the locomotive power of the steamer into our horizon.

Nevertheless, one should not overlook, as Ruskin did in his interpretation of this work, the focus on the spiritual and physical significance of technology in this painting. As Brian Lukacher has clearly identified, the *Snow Storm* should not be misunderstood either as a simple transportation of perceptual and empirical experience or as a mere hallucination. Rather, Turner, who explicitly called himself the picture's "author", emphasised the artist's own responsiveness to the scene as central. Lukacher clearly describes how "the fluid simultaneity of the visionary and the phenomenal" operate within the painting and how elements of "spectral fantasy and poetical allusion" are at work.[17] Turner's pictorial treatment of technology is as central within this work as the phenomenal experience of nature. The painting indeed struggles with the significance of technology, here the steamboat, and its movement through the storm that intensely (through its soot, for example) affects the colour and atmosphere of the seascape. Alluding in the full title of the piece[18] to Ariel, the demon spirit in Shakespeare's *The Tempest*, Turner furthermore leads us to the question of how technology, and the machinery both of technology and of his own painting, is interconnected with the spiritual forces of life.

At the same time as Britain praised enthusiastically the progress of mechanisation as a way out of the economic crisis of the time, the burning question was arising of what technology meant for, and did to, the spirituality and morality of men and women. *Snow Storm* mines this challenge deeply and offers a subtle exploration of it by contrasting the rhythmic motion of the sea with the mechanised locomotive power of the steamer. The agencies of art, nature, and technology are visualised in a masterly way in the work, and one can wonder whether the steam engine also became a model for Turner's own painting machine.[19] If interpreted in this way, Turner's *Snow Storm* formulates a deeply challenging question for our time also: "how can the vestiges of imagination and poetry, the vaporous traces of Ariel, be sustained in a mechanistically-improved environment of locomotive energy and in a commercial visual culture of empirically-minded curiosity?"[20] According to Lukacher, Turner indulges in the contemporary mystification of the machine by reconciling the poetical and the technological, but one need not follow him in this; instead, one can also read the *Snow Storm* as a genius thought-provoking impulse to let the viewer (and reader) him/herself delve into the question of how technical forces and spiritual powers are interrelated. Are ghosts haunting the machine, or are the human-made but now unbound machine demons rather occupying our social world and transforming nature into one gigantic artefact?[21]

Nevertheless, *Snow Storm* serves as a provoking "self-critical social metaphor for the delusive apprehension of reality in an age of steam-propelled Ariels, the image striving to reveal the illusory and mutable forms of the mystifying ideology of technological progress".[22] By overlooking entirely the dimension of technology in Turner's painting, Ruskin in the first place seemed to violate its content, but as Lukacher rightly points out, this should rather be understood as a deeply critical statement against the power of steam as the spirit and *pneuma* of modernity.[23] Ruskin was, in the same intense way as Turner, curious about the effects of weather and industrialisation, about *industrial weather* so to say.[24] But while Ruskin believed industrialisation to be deeply dehumanising,[25] Turner did not regret industrialisation in general. Ruskin distinguished between the clouds of nature and the clouds of industry, while Turner explored just clouds.[26]

Inspired by the entanglement of poetic and natural energy and of technological and spiritual power in Turner's painting, I would like to understand his work as a masterly exploration of the animism of modernity. Was Marx right in proclaiming and complaining about capitalism's victory over traditional animism, or have we rather entered a new time of animating artefacts with and against nature? What kind of a fetishisation is taking place in technology and what kinds of demon spirits are emerging?[27]

Ruskin aptly characterised the artist's work as true to nature. In defending the artist against accusations of his lack of truth and of being "not like nature", Ruskin declared that "Turner is like nature, and paints more of nature than any man who ever lived".[28] Turner was deeply insulted by the muttering criticism of this work as "soapsuds and whitewash", but he nevertheless fortunately scorned this and similar comments. With regard to the expression of weather, his colour beginnings offered him new and unproven tools for expression, which he made full use of. His colour scale, for example, changed more and more along the scale that Goethe had described between negative and positive colour tones, where Turner moved from blue to mainly yellow. In Turner's famous works on the "Deluge", he acknowledges Goethe, whose theory of colour from 1810 had had a profound effect on him (Fig. 2.4).[29] Nevertheless, Turner disagreed with Goethe about the absence of light in darkness, and he found positive values in both darkness and light.[30]

Besides the transparency of light and colour in Turner's vision, the "Deluge" also depicts the human inability to control and reign over nature. One might herein also identify Turner's faith in God as the Giver of Life with an absolute power to destroy but also to create and re-create it. The blurring colours make us aware of how our eyes are involved in the process of re-creating the world. Seeing and feeling become one in the process of re-creation, just as the interplay of light and darkness individually brings about each colour according to Goethe's theory.[31] Turner involves the viewer in this process where yellow constitutes the central light, while the edges get darker as the viewer moves away from the centre (Fig. 2.5).

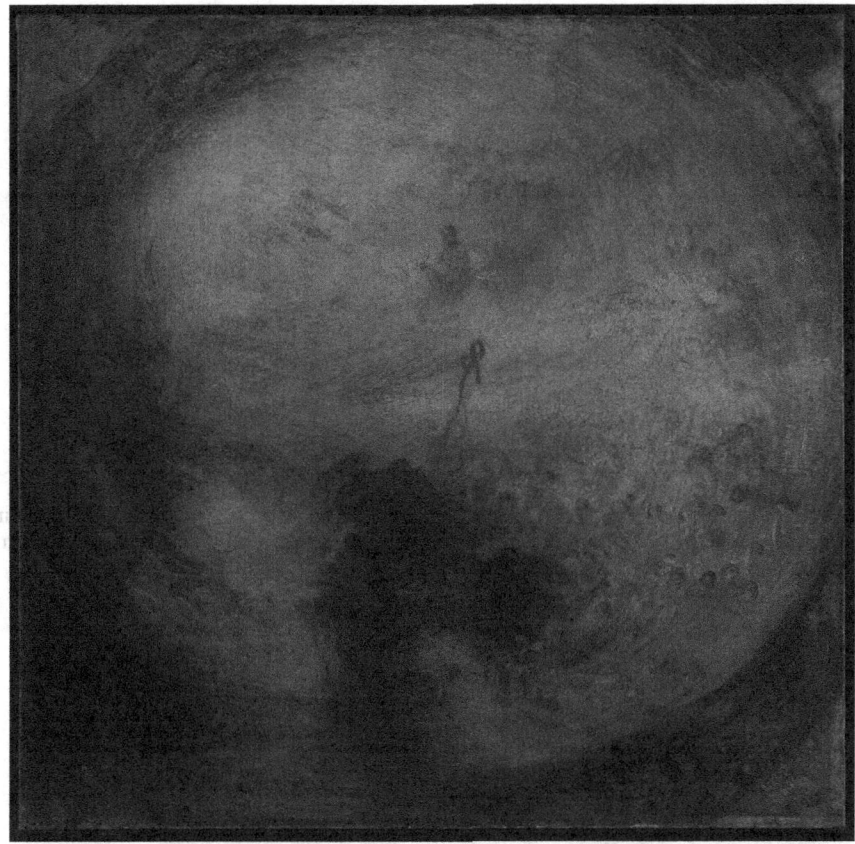

Figure 2.4 Joseph Mallord William Turner (1775–1851), *Light and Colour (Goethe's Theory) – The Morning after the Deluge – Moses Writing the Book of Genesis*, 1843, oil on canvas, 785 × 785 mm, Tate Britain, London.

Photo © Tate

In his sketch book *Liber Studiorum*, Turner includes another version of the deluge, which builds on his larger painting. Both have probably been influenced strongly by Nicolas Poussin's earlier painting from 1660 and Turner's fascination with the scenery's "dismal gloom".[32] The storm and flood waves and the being-exposed-to-the-weather are emphasised even more strongly here. Significantly, it illustrates the power of God within nature and weather according to the biblical story: "Everything on dry land that had the breath of life in its nostrils died" (Genesis 7:22, NIV). Alluding to his *Morning after the Deluge*, Turner again visualises human dependency on God's gifts of life in nature (Fig. 2.6).

Art history usually identifies Turner as a representative of romantic landscape painting and sometimes acknowledges him as a forerunner of impressionism;

Figure 2.5 Joseph Mallord William Turner (1775–1851), *The Deluge*, 1815, watercolour on white wove lightweight writing paper, 204 × 284 mm, Tate Britain, London.

Photo © Tate

Figure 2.6 Joseph Mallord William Turner (1775–1851), *The Deluge*, 1805, oil paint on canvas, 1429 × 2356 mm, Tate Britain, London.

Photo © Tate

one might agree to this latter statement with regard to his colour beginnings in the sketch books rather than with regard to his influence on others, because the impressionists themselves, such as Monet, could not find much of interest in his work. Romantic landscape painting visualises what has been circumscribed as the sublime in nature, meaning the awe-inspiring, powerful, savage, and untameable.

The sublime, furthermore, offered a metaphor for God's power, and we can wonder to what degree light in Turner's painting through the different phases of his work might have "signified the emanation of the spirit of God".[33] If Turner did in fact connect his deep interest in the mystery of light, which runs through all his paintings, to the mystery of the emanation of God's Spirit, we can indeed interpret his weather landscapes in a theological key. Even if one should take seriously what are handed down as his last words uttered before death – "the sun is God"[34] – one should not regard Turner as some kind of a religious painter in the first place.[35] Nevertheless, Ruskin might be right that only Turner could convey "the mysteries of God".[36] His deepest interest was obviously the forces of nature as they manifest in weather on land and at sea. John Ruskin is therefore probably quite right in describing Turner as the artist who could most "stirringly and truthfully measure the moods of Nature".[37] Possibly one might acknowledge Turner furthermore as a pioneer who started to investigate nature and weather within us, as his paintings, as much as they depict the manifestation of nature on the canvas, also explore the manifestation of nature within our process of seeing and feeling. Together with other Romantic painters such as Caspar David Friedrich, Turner also challenged the distance between past and present and between seeing and feeling.[38] Nature acts as a force in both Friedrich's and Turner's paintings; both depict a spiritual synchronicity; both explore the modern subject within nature, even if Turner conserves a stronger continuity with his predecessors.[39]

One of Turner's early paintings visualises his deep interest in the atmosphere created by weather as a dynamic event. In his *Frosty Morning* from 1813 (Fig. 2.7), the painter turned away from depicting the landscape as a map, instead exploring

Figure 2.7 Joseph Mallord William Turner (1775–1851), *Frosty Morning*, 1813, 1137 × 1746 mm, Tate Britain, London.

in colours what weather as a process shift did to the scenery. Graham Reynolds characterises this painting as "an absolute protest against the labelling of land-scape as map-making".[40] To the title Turner added a quotation from James Thomson's "The Seasons": "The rigid hoar frost melts before his beams".[41] By grading white and brown, Turner sensitively portrays how the frost works in the shadows and the mud. The sun is rising but still hidden in the mist; its light is distributed through the light brown, yellow, and green surfaces in the sky. Sharp silhouettes underline the character of the winter in the trees and humans. The painting records a scene from Turner's travels in Yorkshire, and the artist was particularly fond of it. Monet declared that it had been painted with "wide-open eyes".[42]

One of the painter's favourite themes was the sea. Turner, who did not fully fit into his contemporaries' expectations of a gentleman but rather appeared as a rude, and in Constable's view "uncouth", sailor, produced a larger number of maritime paintings, and his first work exhibited at the Royal Academy in 1796 was *Fishermen at Sea* (Fig. 2.8).

The focus on light and waves, which Turner developed more and more deeply over his lifetime, was already convincingly established in this early oil painting. Moonlight is reflected in two ways in two places on the water's surface where both water and the clouds distribute the shimmering. "This painting plays on contrasts of surface and depth just as it plays on the paradox of simultaneous illumination and obscurity", Sarah Monks states in her insightful analysis of Turner's sea-scapes.[43] The subtle interplay of surface and depth is here carefully depicted in the

Figure 2.8 Joseph Mallord William Turner (1775–1851), *Fishermen at Sea*, 1796, oil paint on canvas, 914 × 1222 mm, Tate Britain, London.

Photo © Tate

motion of the wave that carries the fishermen. As humans, they are at the mercy of the elements, water, wind, and light. In the Romantic tradition of his time, the sea was regarded as a testing place for the emotional self, and sea paintings offered excellent conditions for effective aesthetic responsiveness. Often Turner depicts scenes where ships tip towards us, capsizing and sinking. Flotation and the possibility of sinking to greater depths enforce the tension between surface and depth in a way that for Monks obviously also refers to the metaphorical relation of painter and painting, where the sea is "serving not only to buoy up a human world but also as a surface upon which rough, involuntary and inchoate pattern is produced from somewhere beneath, in a manner analogous to the sketchbook and canvas of the modern responsive artist".[44]

Of special interest in Turner's maritime paintings are the waves. In their intense motion they keep alive the tension between surface and depth. Mobility is here not simply transport from one place to another but an eternal back-and-forth movement which again depends on the rhythm of wind and water. In analogy to Turner's intense emphasis on weather alteration as a constant challenge, the force of motion in the sea and the infinite variability of the water's motion was also at the centre of his artistic imagination and sensitivity.[45] Humans are dependent on the waves; they can either threaten to draw the human down to the depths and death or elevate them to heaven. Water surfaces in Turner's work are either dramatically wild or calmly safe, embedding the sailor and viewer in comfort, such as in his Venice landscapes (Fig. 2.9).

Figure 2.9 Joseph Mallord William Turner (1775–1851), *Venice at Sunrise from the Hotel Europa, with the Campanile of San Marco,* 1840, watercolour on paper, 198 × 280 mm, Tate Britain, London.

Could we regard the waves in Turner's paintings as a central metaphor for human existence in environmental dependence on weather as it is revealed for us in the sublime power of winds and waters, in the interplay of light and darkness, depth and surface?

Similar to the "suffering of sea-change" (Monks) and the floating on waves, the storm also carries for Turner an analogous central meaning where natural force stands against culture. Even Hannibal's army passing over the Alps had to face this power, and Turner brilliantly reminds us again in his painting about the power in changing weather, where every hiker in the mountains, similar to the sailor, must be prepared for a quick change from comfortable to threatening weather conditions. Nature implies an unpredictable potential of change, a change from weather as a gift of life to a life-threatening force. Being alive means surviving in such an existential dynamic of change. The artist's challenge is "to store in his mind every change of time and place".[46]

Inspired by Monks' reference to George Bataille, I would like to apply the notion of "alteration" to characterise Turner's view and expression of nature. Alteration is for Bataille, according to Rosalind Krauss, "both decomposition (as in corpses) and the total otherness of the sacred (as in ghosts)";[47] it can be associated with formless subjects and images. Such an alteration seems to take place in Turner's seascapes and "their rapid and alternating movement between different states set in motion".[48] Alteration in Bataille's strict sense – located in a religio-psychologically worked out dialectic of the monstrous and the sacred[49] – is of course not applicable in its original sense to Turner's art, but if we redefine alteration as a method which includes both composition and decomposition and which synthesises the back and forth of the awful-threatening and the life-giving-comfortable, it might work. Similarly to Bataille's definition, alteration thus connotes "a paradoxical movement both upward and downward, a simultaneous movement of gravity and grace"[50] which we can also recognise in Turner's painting as a simultaneity of depth and surface, hope and despair, of seeing, feeling and believing. The often postulated characterisation of Turner as a "pessimist",[51] contrasted for example with C.D. Friedrich as a "melancholic",[52] is in my view inappropriate. Probably it must be blamed on Turner's poem (never found but sometimes quoted in fragments and applied to select paintings) entitled "Fallacies of Hope", which might have emerged from a reading of Thomas Campbell's "Pleasures of Hope", which Turner illustrated in 1835.[53] In my view, rather than a simple negativity, it is the simultaneity of hope and despair and the embeddedness of the one in the other that is at heart in Turner's image universe,[54] while the "fallacy" of hope in those paintings to which he applied fragments of his poem should be interpreted in the context of a political critique of his contemporary "empire-building countrymen warning of the perils that awaited them if they similarly put their own selfish interests above those of the state".[55]

Alteration, in the sense of a simultaneous upwards and downwards movement, would then be a more precise and adequate description of the weather wisdom that is carried through and conveyed in Turner's work. The fluid appears in Turner's art as superior to the alleged steadiness of the earth. What seems to be stable is exposed to continuous imaginative alterations.[56] Weather is not just simply there, it does not simply change from one state to another, but it alternates. Its oscillations are like untameable waves; its swingings are like winds turning into storms and like mighty storms calming. Turner's paintings sensitise us to the dynamic and exciting forces of weather; they awaken in us respect for its overarching power, and they make us aware of our dependence on nature. For Turner – and this might be his legacy to us – this environmental dependence is best expressed in daily belonging to given and alternating weather lands, "roused by every change in nature in every moment".[57]

Light nevertheless is of another quality.[58] In the same way as we in our being alive are dependent on weather conditions, we are in need of light for life, orientation, movement, and cultivation. Seeing is a gift made possible through light that acts in a subtle interplay with darkness. If the sun is God, then God is light, and weather represents God's effective work in and through nature. The Creator God we meet in Turner's art is therefore for me a God revealing him/herself in light and weather. Being alive as God's image therefore means to be alive as a transparent being that needs to increase his/her skills of effective aesthetic response to weather and light. Visual arts and painting receive in such a context the task of an educational and liberating force. Turner's art orchestrates the growth of the seeing body and believing soul in a given environment of natural alteration. The Spirit as a Giver of Life takes place in subtle ecological entanglements where weather atmospheres reveal the Sacred and where light is experienced as the Creator who embraces his/her creation.

Following German art historian Heinz Ohff, Turner should be acknowledged as the one who has indeed *invented* weather.[59] Even if landscapes have been painted intensely in the history of European art, it was late Romantic landscape painting that first explored the entanglement of weather conditions, such as light, humidity, air, and evaporation, with our human inner sensitivities. Masaccio and Bellini in the 15th century were the first to explore landscapes as spaces, and weather first appeared later in the 17th century as a phenomenon in its own right when the Netherlandish painters depicted misty atmospheres, storm clouds over the sea, and dark grey skies. Clouds and the sky were first taken seriously as a meteorological phenomenon at the beginning of the 19th century, due to Howard's and Goethe's emphasis. However, weather appeared at that time mostly as a part of topography, as a way of framing the land and surrounding it from above. Obviously, it was Turner who first established weather as a central visual theme for painting. As we saw previously, his creative mode of painting allowed him to transform the visual landscape from a map into a weather atmosphere.

Ohff discusses why weather appeared so late in the history of art in spite of its central significance to human ecology through the ages. His preliminary answer is that it might be due to the general discomfort of being completely at the mercy of this external power. Does weather, he asks, provoke an experience and consciousness of one's ephemerality which is threatening?[60] The strong and strange reactions to Turner's moving and captivating large paintings, where the power of weather in all its uncontrollability is overwhelming the observer, might support such an explanation. One can wonder if it still is the same feeling of completely being dependent on something that is far out of our power to influence that creates both a sensory stimulus and a disturbing quality of human life that we prefer to suppress rather than to accept. Does weather remind us all too much about life's vulnerability and ephemerality? Does it disturb and question our identity as autocratic beings with the power to do anything? Is this also a contributing cause to so-called climate scepticism where one entirely, and against all scientific evidence, tries to deny the impact of climate change on our social and economic behaviours?

If there is some truth in this, and I think there is, Turner's paintings, and other expressions of the embracing power of weather over our life worlds, carry an essential wisdom that is necessary to cultivate for our future. Living in weather lands then means to accept and not to resist living under uncertain conditions, to respect *the dignity of alteration*, and to rest in the givenness of life and empathise with one another rather than to nourish the illusion of autocracy. Safe shores are no longer in sight, but only flowing light, misty uncertainty, and an atmosphere as if the earth is still in the early stages of its creation. Turner appears as a master of such insight and his paintings manage to implant it deeply under our skin.

In his dispute with Ruskin, Turner appeared as a liberal person far removed from metaphysics. His skill in painting the clouds, his cloudiness, was respected and honoured by Ruskin, but nevertheless interpreted as a pantheistic mode of de-deification. The clouds and other weather elements were now achieving an intrinsic value; they turned into symbols for human beings, life, and existence rather than referring to the divine. Ruskin complains about Turner's "faithlessness". Weather lands turn in his view into a surrogate for the lost gods. As shown earlier, such an interpretation need not necessarily have the last word, and we can without doubt also approach Turner's art as a revelation of the mysteries of nature, wherein the believer recognises Creation and the Spirit as the Giver of Life. In the lens of Turner's images, nature would then reveal a life-giving Spirit who acts in an unpredictable alteration of weather, and as a Giver of life-enhancing, all-embracing, and place-creating light.

For us, Turner can serve as a master of a modern mode of existence where the danger and uncertainty of "life in turmoil", to use Rilke's striking expression, is exposed at its peak. Being alive now means being exposed to a continuous flow of alteration and not possessing any certainties. Weather teaches us to accept being at its mercy. Turner, and especially his colour beginnings, therefore offers us an outstanding place where we can become aware of being within the world and can discover and accept the world with all its power of change within us (Fig. 2.10). Looking at these paintings makes me spiritually and bodily aware of the dramatic power of the gift of life in weather lands.[61] They might be located in the context of an emerging and accelerating modernity, but nevertheless they represent an encounter with the Spirit who gives life, and vivifies ephemeral and vulnerable beings in unpredictably alternating environments.

Figure 2.10 Joseph Mallord William Turner (1775–1851), *Waves Breaking on a Lee Shore at Margate (Study for "Rockets and Blue Lights")*, ca. 1840, oil paint on canvas, 597 × 952 mm, Tate Britain, London.

Photo © Tate

Notes

1 For a survey of Turner's life and work, see Graham Reynolds, *Turner*, London: Thames & Hudson 1969, reprint 2000, and Martin Butlin, "Rain, Steam, and Speed," in: Evelyn Joll, Martin Butlin, and Luke Herrmann (eds.), *The Oxford Companion to J.M.W. Turner*, Oxford: Oxford University Press 2001. The Tate's website in the context of the recent exhibition "The EY Exhibition: Late Turner – Painting Set Free," 10 September 2014–25 January 2015, also serves as a good introduction to the artist's work: www.tate.org.uk/whats-on/tate-britain/exhibition/ey-exhibition-late-turner-painting-set-free/introduction-to-turner, accessed 27 January 2015. Mike Leigh's film *Mr. Turner*, released on 31 October 2014, offers an exciting portrait of the painter's biography but suffers in my view from a failure to connect this to the paintings and a not unusual asymmetry between his life and his work.

2 In a lecture, Turner explicitly reflected on the advantages of the British climate for landscape art: "In our variable climate where [all] the seasons are recognizable in one day, where all the vapoury turbulence involves the face of things, where nature seems to sport in all her dignity and dispensing incidents for the artist's study . . . how happily is the landscape painter situated, how roused by every change in nature in every moment, that allows no languor even in her effects which she places before him, and demands most peremptorily every moment his admiration and investigation, to store his mind with every change of time and place". Andrew Wilton, *The Life and Work of J. M. W. Turner*, London: Academy Editions 1979, quoted in John E. Thornes, "A Brief History of Weather in European Landscape Art," *Weather* 55, October 2000, 363–375, 367f.

3 Eberhard Roters, *Jenseits von Arkadien: Die romantische Landschaft*, Köln: DuMont 1995, 57.

4 For the theory of atmospheres cf. Gernot Böhme, *Atmosphäre: Essays zur neuen Ästhetik*, Frankfurt am Main: Suhrkamp 1995; and Sigurd Bergmann, "Atmospheres of Synergy: Towards an Eco-Theological Aesth/Ethics of Space," *Ecotheology: The Journal of Religion, Nature and the Environment* 11, 3, 2006, 327–357.

5 On Turner's fondness for catastrophes, see Roters, op. cit., 57–60.

6 Reynolds, op. cit., 191.

7 Ibid., 174.

8 Oskar Bätschmann, *Entfernung der Natur: Landschaftsmalerei 1750–1920*, Köln: DuMont 1989, 111.

9 Ibid., 112.

10 Reynolds, op. cit., 197; Butlin, op. cit., 254. In the lower right corner of the painting, one can also glimpse a running hare that might symbolise speed or hint at the limits of technology or even the fear and danger they bring.

11 The term "colour beginnings" stems not from Turner, who entitled them "beginnings", but from A.J. Finberg in his 1909 Inventory of the Bequest. Shanes regards Turner's Beginnings not as expression of an interest in abstraction but as test sheets and experiments with technical methods in sketching and water colour. From 1818 onwards, Turner shifted his focus from tone to colour and separated colour from form, and his Beginnings obviously played a central role in this evolutionary process. Eric Shanes, "Beginnings," in: Martin Butlin and Luke Herrmann (eds.), *The Oxford Companion to J.M.W. Turner*, Oxford: Oxford University Press 2001, 21–23.

12 "1836–47 Modern Painter," in: David Blayney Brown (ed.), *J.M.W. Turner: Sketchbooks, Drawings and Watercolours*, December 2012, www.tate.org.uk/art/ research-publications/jmw-turner/1836-47-modern-painter-r1130133, accessed 27 January 2015.

13 Reynolds, op. cit., 149.

14 In Turner's own words, according to Reynolds, op. cit., 190. In the catalogue for the Academy's exhibition 1842 Turner wrote, *"Snowstorm – steam-boat off a harbour's mouth making signals in shallow water, and going by the lead. The author was in this storm on the night the Ariel left Harwich"*.

15 Cf. Seibold, who analyses Turner's navigating in between abstraction and naturalism, considering the detailed weather phenomena depicted in his works. His meteorological observations, for example of volcanic dust in the stratosphere, are deeply integrated into his paintings' iconography: Seibold's paraphrasing of Turner therefore as a painter of weather is striking. Ursula Seibold, "Meteorology in Turner's Paintings," *Interdisciplinary Science Reviews* 15, 1, 1990, 77–86, doi: 10.1179/030801890789797707. Large volcanic eruptions took place during Turner's lifetime and might have influenced his painting, especially through the experience of a specific twilight that was caused by the dust in the air. Cf. Götz Hoeppe, "Himmelslicht: Spiegelbild des Erdklimas," *fundiert: Das Wissenschaftsmagazin der FU Berlin*, www.elfenbeinturm.net/ archiv/2003/07.html, accessed 11 February 2015.

16 Roters, op. cit., 64f. "Die göttliche Allmacht verkörpert sich im Wetter selbst, an die Stelle der anekdotischen Schilderung vom Einwirken Gottes auf das Menschenschicksal ist die meteorologische getreten, in der das Wehen der Natur als der Mantel einer unbegreiflichen und unvorhersehbaren Schicksalsbewegung faßbar wird, der wir hilflos ausgeliefert sind".

17 Brian Lukacher, "Turner's Ghost in the Machine: Technology, Textuality, and the 1842 *Snow Storm*," *Word & Image* 6, 2, 1990, 119–137, 120.

18 See Reynolds, op. cit.

19 Lukacher, op. cit., 125 continuing a line of interpretation started by Michel Serres and others.

20 Lukacher, op. cit., 135.
21 On the history of demonic machines taking over the world, see Szerszynski's gripping and provocative narration. Bron Szerszynski, "The Twilight of the Machines," in: Celia Deane-Drummond, Sigurd Bergmann, and Bronislaw Szerszynski (eds.), *Technofutures, Nature and the Sacred: Transdisciplinary Perspectives*, Farnham: Ashgate 2015.
22 Lukacher, op. cit., 136.
23 Ibid., 137.
24 Cf. Chapter 6 on the commodification of weather, where we will spin this yarn further.
25 John Ruskin, *The Storm Cloud of the Nineteenth Century*, Orpington: George Allen 1884, 55–63.
26 Jonathan Hill, *A Landscape of Architecture, History and Fiction*, London and New York: Routledge 2015, 103.
27 Cf. Sigurd Bergmann, "'Millions of Machines are Already Roaring': Fetishized Technology Encountered by the Life-Giving Spirit," in: *Technofutures, Nature and the Sacred*.
28 John Ruskin in his preface to *Modern Painters Volume 1*, 1844, "Preface to the Third Edition," in: E.T. Cook and Alexander Wedderburn (eds.), *The Complete Works of John Ruskin, Library Edition the Works of John Ruskin*, Vol. 52, London: George Allen and New York: Longmans, Green and Co 1903, www.lancaster.ac.uk/depts/ruskinlib/Modern%20Painters, accessed 28 January 2015.
29 Reynolds, op. cit., 194.
30 Gerald Finley, *Angel in the Sun: Turner's Vision of History*, Québec: McGill-Queen's Press 1999, 202.
31 Both pictures of the Deluge were stolen from the Schirn Kunsthalle in Frankfurt/M. in 1994. On Turner's adaptation of Goethe's colour theory, see John Gage, "Goethe, Johann Wolfgang von," in: Martin Butlin and Luke Herrmann (eds.), *The Oxford Companion to J.M.W. Turner*, Oxford: Oxford University Press 2001, 127f.
32 Matthew Imms, "The Deluge c.1815 by Joseph Mallord William Turner', Catalogue Entry, May 2006," in: David Blayney Brown (ed.), *J.M.W. Turner: Sketchbooks, Drawings and Watercolours*, January 2015, www.tate.org.uk/art/research-publications/jmw-turner/joseph-mallord-william-turner-the-deluge-r1154528, accessed 26 January 2015.
33 http://hoocher.com/Joseph_William_Turner/Joseph_William_Turner.htm, accessed 26 January 2015.
34 Norman Davies, *Europe: A History*, London: Pimlico 1997, 687.
35 In contrast to a common belief that Turner was religiously disinterested, we can find traces of the opposite, as he seems to have followed his two friends and ordained ministers, Trimmer and Daniell, with true interest. The sketch books include evidence for Turner's religious feelings and also church attendance. Hamilton therefore wonders rightly whether "religious sensibilities may have been the driving force behind such paintings as *Angels standing in the Sun* and *Light and Colour*". James Hamilton, "Private Life," in: Martin Butlin and Luke Herrmann (eds.), *The Oxford Companion to J.M.W. Turner*, Oxford: Oxford University Press 2001, 240–244.
36 Ruskin, *Modern Painters Volume 3*, 301: "With him, the hue is a beautiful auxiliary in working out the great impression to be conveyed, but is not the chief source of that impression; it is little more than a visible melody, given to raise and assist the mind in the reception of nobler ideas – as sacred passages of sweet sound, to prepare the feelings for the reading of the mysteries of God". Cf. Michael Wheeler, "Environment and Apocalypse," in: Michael Wheeler (ed.), *Ruskin and Environment: The Storm-cloud of the Nineteenth Century*, Manchester: Manchester University Press 1995, 173.
37 See David Piper, *The Ilustrated History of Art*, London: Chancellor Press 2000, 321.

38 Sarah Monks, " 'Suffer a Sea-Change': Turner, Painting, Drowning," *Tate Papers* Issue 14, 1 October 2010, 3, www.tate.org.uk/research/publications/tate-papers/suffer-sea-change-turner-painting-drowning, accessed 28 January 2015.

39 For an insightful comparison of C.D. Friedrich and J.M.W. Turner, see Roters, op. cit., 61–65.

40 Cf. Reynolds, op. cit., 89.

41 James Thomson, *The Seasons*, 1726–30. *Autumn*, 165, in: James Thomson, *The Seasons*, with His Life, an Index, and Glossary and Notes to *The Seasons* by Percival Stockdale, T. Chapman 1793, https://archive.org/details/seasonsbyjamest00thomgoog, accessed 27 January 2015. Probably the whole of the following section of the poem inspired Turner's painting:

> "The lengthened night elapsed, the morning shines
> Serene, in all her dewy beauty bright,
> Unfolding fair the last autumnal day.
> And now the mounting sun dispels the fog;
> The rigid hoar-frost melts before his beam;
> And, hung on every spray, on every blade
> Of grass, the myriad dew-drops twinkle round."

42 www.tate.org.uk/art/artworks/turner-frosty-morning-n00492, accessed 27 January 2015.

43 Monks, op. cit., 3.

44 Ibid., 4.

45 Cf. Luke Herrmann, "Waves," in: Martin Butlin and Luke Herrmann (eds.), *The Oxford Companion to J.M.W. Turner*, Oxford: Oxford University Press 2001, 373f.

46 Cf. Wilton, op. cit.

47 On Bataille's concept of alteration, see Rosalind E. Krauss, *Bachelors*, Cambridge, MA: MIT Press 1999, 7f; Cf. Rosalind E. Krauss, *The Originality of the Avant-garde and Other Modernist Myths*, Cambridge, MA: MIT Press 1986, 54.

48 Monks, op. cit.

49 Jeremy Biles, *Ecce Monstrum*: *Georges Bataille and the Sacrifice of Form*, New York: Fordham University Press 2007.

50 Ibid., 94.

51 Eric Shanes, *Turner*, 18, www.turnersociety.org.uk/Turner_biography.pdf, accessed 28 January 2015. Kenneth Clark, *The Romantic Rebellion*, New York: Harper and Row 1973, 236f., emphasises Turner's "deep pessimism", but in my view this is applicable rather with regard to his political position in the wars and empire conflicts of his time than to his existential exploration of nature and humankind. Clark, op. cit.

52 Roters, op. cit., 61.

53 www.tate.org.uk/art/artworks/turner-the-andes-illustration-to-the-pleasures-of-hope-tw0181 and www.tate.org.uk/art/artworks/turner-kosciusko-illustration-to-the-plea sures-of-hope-tw0182, accessed 28 January 2015.

54 The double attitude of pessimism and optimism is also visible in Turner's two paintings of the Deluge, where *Shade and Darkness – The Evening of the Deluge* expresses his pessimism about the weakness of humanity while the morning picture *Light and Colour – The Morning after the Deluge* breathes optimism and expresses Turner's hope for human and artistic creativity that "can give new life to mankind". Mordechai Omer, "Biblical Subjects," in: Martin Butlin and Luke Herrmann (eds.), *The Oxford Companion to J.M.W. Turner*, Oxford: Oxford University Press 2001, 23f.

55 Eric Shanes, "Chairman of the Turner Society, the Life and Art of J. M. W. Turner, RA," www.turnersociety.org.uk/Updated_biography_life.pdf, accessed 28 January 2015, 36. The same strongly political criticism of imperial power is also carried by

his "The Sun of Venice Going to Sea" from 1843. According to his never seen poem, a "demon in grim repose" lay in wait for the boat. Turner depicts the sun on the sail, usually symbolising the Venetian rulers' power in subtle contrast to the sunlight that shines on a still, calm water surface where one nevertheless can intuit the fatal storm. For a detailed analysis of the sun in this work, see Joseph Hillis Miller, *Illustration*, Cambridge, MA: Harvard University Press 1992, 147–149.

56 Roters, op. cit., 53f.

57 Cf. Wilton, op. cit.

58 Cf. the following detailed study on light in Turner's work: Ursula Seibold, *Zum Verständnis des Lichts in der Malerei J. M. W. Turners*, Dissertation, Heidelberg 1987.

59 Heinz Ohff, *William Turner: Die Entdeckung des Wetters*, München: Piper 1987. Roters, op. cit., 53, also ranks Richard Parkes Bonington and John Constable with these but agrees with Ohff on the pioneering achievement of Turner.

60 Ohff, op. cit., 15.

61 Turner continues to inspire modern artists in a remarkable way, where his emphasis on the exploration of colour and light seems to be especially significant. Danish-Icelandic artist Olafur Eliasson, for example, departed from a selection of Turner's paintings in order to analyse the pigments and isolate his colours, and thereby achieve a light spectrum in his "Turner colour experiments", displayed at the Tate in 2014–2015. Eliasson emphasises Turner's skill in establishing ephemeral effects and traces these within his own long-term Weather project. Cf. Olafur Eliasson, "Reality Is Ephemeral: Turner Colour Experiments," www.tate.org.uk/context-comment/articles/reality-ephemeral, accessed 27 January 2015.

Cf. also Mark Rothko's deep affinity with Turner explored in the exhibition "Turner/Rothko" at the Tate in 2009: www.tate.org.uk/about/press-office/press-releases/bp-british-art-displays-turnerrothko, accessed 27 January 2015. Rothko felt a deep kinship with Turner – whom he claimed had "learnt a lot from me" – and reflected explicitly on the connection between abstract expressionism and the sublime in his influential text "The Sublime Is Now" from 1949.

On Turner's influence on the French impressionists, mainly Claude Monet and Camille Pissaro, see Allen Staley, "Influence on Later Artists," in: Martin Butlin and Luke Herrmann (eds.), *The Oxford Companion to J.M.W. Turner*, Oxford: Oxford University Press 2001, and "Posthumous Reception Abroad," in: *The Oxford Companion to J.M.W. Turner*, 236–238.

3 Atmospheres agog

Weather, culture, religion

A central point in the previous chapter was to interpret the painter as a master of a modern mode of existence where insecurity and uncertainty are exposed at their peak, and where the exploration of weather through observing and painting serves to express wisdom about the existence of volatile and vulnerable beings in unpredictably alternating environments in the context of emerging modernity. Regarding Turner's painting as "inventing weather" – in the sense that being alive means being exposed to a continuous flow of alteration over which we have no control – does not of course imply that weather did not have an impact on human life before. Humans were struggling with weather well before Turner, teaching us to accept being at its mercy.

The kind of wisdom that recognises how vulnerable humans and communities are at the mercy of the weather certainly also developed in earlier times and in other places than Turner's early industrialising England. This chapter will collect observations from different world regions and times about what we might call weather images and weather responses. How have people perceived and imagined weather in different contexts and how did they respond? In seeking to answer these questions, it will draw on selected secondary sources in the fields of cultural studies and religious studies, where one certainly can trace our theme of weather but where systematic investigations into the reciprocal impact of weather and culture/religion are still a lacuna.

This applies with one exception. James Frazer's extensive description and discussion of "the magical control of the weather" and "the magical control of rain" in two sections of his *The Golden Bough*, a pioneering study in the history of religion, is certainly wide ranging and rich in context and details, but one should take it with a grain of salt, as it is impossible to decide to what degree the reports build on the author's sources and to what degree on his powerful literary imagination.[1] Frazer's book title, it should be noted, is taken from a famous landscape painting by Turner.[2] The painting stems from the imagery of the sacred grove inspired by Virgil's poem, the *Aeneid*, where the Sibyl cuts a golden bough from a sacred tree to allow the Trojan hero Aeneas to enter the Underworld. In this image, Frazer sees his central thesis illustrated – for in it, "the whole system of mythology" and "the evolution of religion" are visualised in the "plucking of a bough – the Golden Bough – from a tree in the sacred grove"[3] by the priest, the King of the wood – himself an incarnation of a dying and reviving god. Frazer's theory of religion alludes to Turner's tree in a sacred grove pregnant with meaning. But while Turner's work emerges in comfortable weather and warm Italian light, Frazer's book takes another course and instead focuses on the darkness of human belief and the violence of religious practice.

Even if contemporary studies of religion do not need to follow Frazer's theses, or agree on his characterisation of magic, or his idea of an evolution from magic through religious belief to scientific thought, his exposition of a broad range of primary and secondary sources can still offer inspiration. Nevertheless, Frazer, in accordance with his overarching framework, regards magic as a misleading superstition prior to the advent of religion and science, an attitude and prejudice that Ludwig Wittgenstein with good reason sharply criticised in his *Remarks on The Golden Bough*: "it makes these views look like errors".[4]

Some other informative surveys of "weather magic" can also be found, even if these mostly build on ancient written and secondary sources such as Ádám Molnár's *Weather Magic in Inner Asia*.[5] Only a few analytical explorations of the interconnection of weather and religion have appeared on my bibliographic radar.[6] The intention of this chapter, therefore, is by no means to offer a comprehensive survey of the subtle interaction of weather, culture, and religion but to enrich the elaboration with short excursions and insights into other geographical, historical, cultural, and religious contexts.

In the chapter's concluding reflection on this journey of glimpses into selected religious weather responses, the notion of "atmosphere" will offer us an important analytic tool and a central thread that the book's final chapter will also spin further. While "atmosphere" represents a central term in the science of meteorology, it will here be used as a hermeneutical lens in the phenomenological sense. Culture, and religion within it, responds to weather as atmospheres that are taking place *in between* – between culture and nature, between humans and gods, and between spirits and natural elements. Weather, one might say, produces atmospheres of belief in culture. My short panorama in the following can thus only serve as an indication of the surprisingly rich multiplicity of different symbolic imaginations of weather phenomena that appears on the map of diverse cultural and religious traditions. One can only hope that it will inspire other scholars in religious and cultural studies to mine more deeply and systematise this exciting field.

Fiji – biocosmic religion and climate justice in Pacific storms

The journey starts in the Fiji islands in the Pacific, a land of alleged paradisiac weather conditions with a comfortable uniform average temperature of 20–30°C on the one hand and life-threatening storms and hurricanes, floods, and droughts on the other, seriously disturbing the picture of a paradise on Earth.

Observations about Fiji offer the point of departure for Simon D. Donner's introduction to a volume of the influential journal *Climatic Change*, which explicitly unfolds the groundbreaking argument that climate science needs to be completed with studies from the humanities, especially in the thematic field of religion.[7] Donner offers rich insight into how inhabitants of the Fijian village of Matacawlevu regard their agriculture, physically and spiritually, intimately connected to both weather and the cultivated soil. The villagers celebrate annually the planting of their crops, and approach the weather ritually in a rich mix of animism and Christian Methodism. The rain that is needed for good growth and a plentiful harvest is assured by the prayers in the ritual, and potential droughts are connected to an alleged lack of devotedness. According to Donner, the tradition of the planting festival masks "a sophisticated system of land management that has sustained indigenous people in Fiji and across many islands in Polynesia and Melanesia for centuries".[8]

Interestingly, the people in the Fijian village differentiate strictly between land and weather. Through their farming practices they have some control over land-based aspects of agriculture (such as sowing, watering, and harvesting), while the weather is outside of their influence and power and therefore needs to be approached ritually. The village's agricultural practices follow directly from a deeply held belief that people exert control over land. The weather, however, is up to God.[9]

The notion of "weather land" that was crucial in the introductory chapter changes its meaning slightly in Fiji. While the land belongs to the people – or better, while the people belong to the land – the weather belongs to the Divine – that is, in the context of Christian mission, God the Creator. God impacts on human ecology through weather that affects the fertility of the soil. Humans can execute power over land, but only God can manage rains, droughts, and storms. Therefore, good use of the land also necessarily includes the prayers and rituals for weather that enhances the community's life and survival.

I am not as certain as Donner that the domains of sky and land, of God and human, are so strictly separated as he claims. In both traditional religion and the Christian faith, spiritual forces and God are always present and active on Earth as well as beyond and above; the commodification of the "sky" as a domain for God is a modern Western invention and does not fit into either a biblical or a historical Christian worldview or the kind of pre-Christian traditional religious view found in Oceania. One must, nevertheless, agree with Donner asking whether this common belief – that God controls the sky and the weather while humans are in control of the land – represents a crucial obstacle to educating the people of Fiji about the anthropogenic effects of ongoing climate change, where rising sea levels, dangerous storms, and floods are by no means an act of God but a result of economy-driven human processes.[10]

Of interest, however, for our discussion here is the obvious entanglement of weather and land. Weather, understood as flux along the same lines as discussed earlier, impacts on and penetrates land as fertile soil for agriculture and the human community's life. For the sake of the land, weather needs to be religiously and ritually approached. Their attitude of reverence is evident for weather impacting on land, and the underlying perception of the environment and the variety of weather as a gift – a gift that by no means only enhances but also threatens life and therefore must be ritually regulated. The flux of weather lies in the hands of God.

Faith is an integral part of human ecology. Weather affects land, land affects people, and people praying to God affect weather that again affects land and people. God, weather, land, and people have come full circle.

Ongoing climate change, however, crucially disturbs and even damages this circle;[11] it threatens not only the weather islands of Oceania but also the spiritual universe of its inhabitants. Is the full circle thrown into disarray when men/women take the place of God and affect the weather? Rituals and prayers represent a mode of communication between the God of Weather and the humans on Earth, but to whom should one pray now and for what?

While prayer in the older weather (is)lands of Fiji was able to regulate symbolically the interaction of land, agriculture, and weather flux, and also assisted the constructive adaptation of the farming villages to the continuous flow of weather variation, climate change today not only disorders the physical pattern of variation but also impacts radically on the cultural and spiritual perception of the environment and oneself within it. Climate change science and corresponding national as well as international geopolitics are at present turning into arenas that earlier were manifested by festivals, rituals, and prayers. What turn is religion taking in this challenging situation? In this context, prayer addressed to God can rarely be only for life-enhancing weather directly; should it instead be for the humans affecting the weather, and for mitigating climate change?

Such a mode of believing and thinking is obviously in dynamic development in the Fiji islands and seems to have substantial roots in their history. Ancient religion in Fiji involved prayers and rituals, as well as sacrifices, in order to prevent storms and droughts. It interpreted the failure of annual rain and floods as punishment for sins, and thus also attributed to the role of the shaman earlier and the priest later great significance as mediator. Although they mainly relied on the rituals, societies may also have included individuals who possessed specific magical powers to command the weather.[12] Fruitful seasons, fine weather, and suitable winds for sailing, but also earth tremors, could be assigned to the spirits, who always had to be consulted. With the term "mana" one circumscribed the immanent force of life, arbitrary and uncontrollable, a true force that has come to pass. The missionary Thomas Williams, in his ethnography from 1858, characterised the islanders' religion as belief in "an invisible superhuman power, controlling or influencing all earthly things".[13]

Along with the full circle of weather, land, people, and God described earlier, such a traditional image of weather and response to it scarcely makes sense any more. Regarding weather through the lens of climate science makes it impossible to simply regard weather as "an act of God" (a term that is still used in the language of modern insurance companies!) but renders it necessary to explore the conflict between traditional religious and biblical Christian views of weather and dramatic sea level rises, increasing stronger storms, and challenging droughts and floods caused by anthropogenic economic activity in other parts of the world. Does this mean that God, the Creator, has lost control of the weather? And if so, has he/she also lost control of the "images of God", the humans? Or the opposite: is God angry with humans and enacting weather as a punishment?

Neither of these two paths really seems to fit well into traditional beliefs or Christian faith.

A third path is taken by the Pacific Church leaders in the Moana declaration from 2009, where the issue of resettlement is taken seriously and the rights of the populations are claimed to be ensured. This leads the church to

> *[r]eaffirm* the prophetic role of the church and its responsibility to recognize and speak out against the injustices wrought on by climate change and call on all persons, communities and states to act now.[14]

The churches in the Pacific today, instead of emphasising the anthropocentric Unity-in-Christ-thinking, stress the notion of "the household of God" in their ecumenical narrative, and accentuate relationality in their interpretation of life within church and society.[15] Weather in this way moves from the hands of God into the hands of humans in synergy with the Creator, and the ritual care for the environment turns into a prophetic commitment to act in synergy with God for social and environmental justice both on the local and on the geopolitical global scene. Weather atmospheres in between land and sky and responses to weather flux are turning into political negotiations between nations, companies, and the populace, which again is divided between rich and poor, protected and suffering parts of the world population. Weather is no longer in the hands either of God or of praying local people – so whose hands is it in?

According to John D'Arcy May, the religious centre of the whole of Pacific culture is the all-embracing life.[16] The central spiritual force is the breath of life, and social order mirrors cosmic order. Social relations and cosmic relations are entangled and potential disharmony with the forces of nature always has a personal or social root. That is also why weather has to be approached in a religious way. Analytically, D'Arcy May differentiates between "biocosmic" and "metacosmic" religion in the Pacific. While metacosmic religion offers salvation and a way out of suffering and the conditions of transient life, biocosmic religion is centred on the cosmos and the life that emerges from it.[17] In the traditional worldview one does not differentiate between religious and non-religious experience but rather values being in relationship with the forces of a universe exploding with life.[18]

Even if Christian mission to the Pacific, and especially to Fiji, has also implied evangelical condemning attitudes and negative cultural impact, D'Arcy May describes hopefully the contemporary Christian path to an ecumenical future where Christians, Hindus, and Muslims, who are represented on the islands, develop one common cultural and political community and where belief in Christ develops as faith in "Christus Integrator", the one who heals and unites for the sake of all.[19] Together with the vision of the church leaders for a prophetic church that acts for environmental and social justice, such an ecumenical path might offer the soil for a new weather image and for new practical responses to the flux of weather, including in times of dangerous climate change.

In their politics of recent years, leaders of the Fiji islands have so far not viewed migration or resettlement from the islands as a priority. Instead, they focus on environmental damage and the resulting decrease in living standards of the population. Maybe Fiji will even become a host for other displaced persons and climate refugees.[20]

As I am writing this part of the chapter, the Fijians have just been hit by the strongest storm in their history, and they are still struggling acutely with coming to terms with its damage. Tropical cyclone "Winston" smashed into the islands on 20 February 2016, killed 43 people, and levelled entire communities. Tens of thousands of people are lacking food, water, and shelter. Villages have been wiped out, crops have been uprooted, and infrastructure lies in ruins. Approximately 40% of Fiji's population, 347,000 people, is affected, whereof, according to UNICEF, about 120,000 are children. The storm is regarded as the most powerful storm to have ever made landfall in the southern hemisphere. In the context of global warming and the widespread effects of El Niño, one might expect further similar disasters in the future.

Ahead of the talks of the international climate change conference, COP 21, in Paris in December 2015, Fijian Prime Minister Frank Bainimarama warned,

> Unless the world acts decisively in the coming weeks to begin addressing the greatest challenge of our age, then the Pacific, as we know it, is doomed.[21]

Cyclone Winston and its comprehensive nation-wide destruction might serve other nations as a harbinger of what climate change might bring in "our common future". For faith communities and religions, it might serve as an even stronger catalyst to pray, hope, and act politically for global climate justice. One of the tales collected by Australian media in the storm's aftermath summarises this catalysing force; church elder Eleni Tinai tells the journalist,

> In one of the churches they sang and prayed for better days. . . . But there was no choir at the Seventh Day Adventist Church, which has literally gone with the wind. . . .
> We lose our hope when we saw our churches damaged. . . .
> For us, we have to take hands together and build up a new church for our community.

Before leaving the Fiji islands, let us take a last look at the art of building. Considering the weather lands of Fiji also leads to the question of how weather alteration is encountered in the country's traditional architecture. Even if the strength and capacity for damage of storms is increasing in the context of global warming, storms and hurricanes belong to the natural weather variation on the islands. Traditional architecture here and in the whole Pacific therefore had to adapt to highly difficult conditions, where one often needed to rebuild houses after storms, and where one even tried to build in a way that could minimise the destructive effects of a storm. Traditional houses were therefore constructed without walls but with woven blinds that could be lowered for harsher weather. For the same reason, Fijian houses often had (and still have) walls and roofs of reeds or woven palms.

For religious and social purposes, every village had a common meeting house and a "spirit house" (or "bure") (Fig. 3.1). Before the arrival of Christianity, each post of a bure that was raised required human sacrifices, sometimes even the life of the craftsmen.

Large ceremonial buildings for religious functions are still constructed in this traditional weather-appropriate fashion. Architecture, we may say, cooperates fruitfully with culture and religion in a specific life-enhancing synergy in the Fiji weather (is)lands.

It is highly probable that the dynamic reconstruction of vernacular architecture, where building styles are able to adapt to environmental hazards, can offer an important strategy for coping with the effects of dangerous and increasing climate change. This might also serve as a reminder for sacred architecture, where Christian churches are stone-built in massive European and colonial styles, while Hindu temples, although also built in stone, more explicitly draw on local vernacular architecture.[23]

In accordance with what the missionary once perceived as a "taste of fine arts" and perfection of building skills along the view cited by Thomas Williams in 1858, "in architecture the Fijians have made no mean progress",[24] the progressive

Figure 3.1 Bure of Na Ututu, a sketch made in the early 1800s.[22]

https://commons.wikimedia.org/wiki/File:Bure_of_Na_Utuutu_(Boston._Congregational_Publishing_Society,_1871).jpg, accessed 29 January 2020

architecture for the future might be precisely of this style, vernacular and soundly adapted for weather lands, a thread that we will take up again in Chapter 7.

Priapos – sexual power, fertility, robbery, and gendered weather in ancient times

While weather religion in Fiji seems to have followed a pattern where the spirits, in ancient times, and the Christian Creator God, in later times, have been approached as masters of weather, and where beliefs and rituals have aimed at a pattern of weather variation that sustains human life, religion in the culture of Roman and Greek antiquity offers a different embedding of the religious perception of weather, which it connects to fertility, sexuality, and eroticism.

In the classical world, sexuality represented the struggle between everyday and divine life; sexuality had to do with the fear of death and the continuation of life. With regard to beliefs in classical weather lands, the Greek god Priapos (Latin Priapus) appears as a complex divinity, deserving of a closer look through the lens of our theme.[25] One of the most popular deities in ancient Rome, he was a minor god of fertility, and protector of crops, but also responsible for diseases, robbery, and bad weather.[26] In Fiji, we saw the divine oversight of the connection between agriculture and the fertility of the soil on the one hand and the significance of weather variation on the other, but in antiquity, the adoration of the god(s) impacted not only on this connection but also on human and social aspects such as sexuality, health, and crime. Bad weather, bad behaviour, and bad social habits seem to form a fellowship that one could approach with the help of Priapos.

The intimate triangular connection between the fertility of the soil, where life-sustaining crops are produced, the fertility and sexuality of humans, and the involvement of deities is well known from different rural cultures, but in antiquity the social aspect gained a greater significance, where bad weather and bad behaviour have something in common and where both are rooted in the god's existence. Priapos was well known as the protector of gardens, and on mosaics, artefacts, and paintings he is depicted with a large phallus, which served as a symbol for virile sexual power as well as a symbol of fortune with regard to the cultivation of gardens that needed adequate weather conditions (Fig. 3.2). Some historians regard the phallus of Priapos as the oldest original design for the first cultivated garden, but one might doubt. Until the 19th century, sculptures with Priapos and his erected phallus were quite common across Europe.

The relation between weather, garden, and robbery obviously has its roots in the role of Priapos as protector of the garden, guarding it against thieves who would steal its fruit. In the *Carmina Priapea* (from the 1st century AD or the beginning of the 2nd century), the god wonders why his property is exposed so often to robbery. In the epigrams, Priapus frequently threatens sexual assault against potential thieves, and sculptures of the god were therefore used to protect gardens and cultivations from thieves. In line with his protecting function, Priapos was also regarded as the god of punishment, a punishment that was appropriate to the gender and age of the perpetrator and executed sexually in a shameful way.

Figure 3.2 Priapus with Double Phallus, fresco from the Lupanar in Pompeii, north wall, between rooms c and d, ca. 70–79 AD

Photo: Wolfgang Rieger, in John R. Clarke, *Ars Erotica*, Darmstadt: Primus 2009. https://commons. wikimedia.org/wiki/File:Pompeii_-_Lupanar_-_Priapus.jpg, accessed 22 August 2019

A garden in modern times that alludes to the ancient adoration of Priapos was created in 2009 in the small Swedish city of Gävle, where artist and scholar Ingo Vetter launched his "Garden of Priapos" in the middle of one of the city's three artistically designed roundabouts, "trafikplats Gustavsbro" (Fig. 3.3).[27] A vertically oriented large pink car was placed in the centre and surrounded by green beds, bushes, and trees. Priapos' phallus was here represented by the upward-rising, erect car, an old Volvo jeep, painted pink, a symbol of contemporary maleness, and the original form of all gardens was transferred straight into the world of environmentally damaging motorised mobility.

Sculptures of Priapos in older times were also commonly found at the doors of buildings and at crossroads, and the modern Swedish Garden of Priapos continues this ancient tradition. Inhabitants' reactions to the garden were ambivalent, to put it mildly. In 2017, the sculpture was vandalised. Unfortunately, the weather dimension of Priapos' power is not represented in the artwork, unless one identifies it in the simple need for adequate weather for the garden's plants to grow and flourish, and for the drivers in demanding Scandinavian weather conditions to navigate the roundabout safely. I leave it to the reader to judge how far one

Figure 3.3 Ingo Vetter, *Priapos Trädgård*, Gävle, 2009.

Photo © Ingo Vetter, by courtesy of the artist. http://ingovetter.com/project/priapos-tradgard/, accessed 22 August 2019

might appreciate this late modern Priapos and his power to keep bad weather at a distance. The location of Priapos at the entrance to the city qualifies him as a guardian, and it should be Priapos who watches over and guards the gardens of the city of Gävle. An erect car symbolises clearly male sexuality and the car as man's elongation. The car also serves as a symbol for serious environmental damage, and can at the same time relate to other narratives and works of art in the city.[28] Will the god thereby also protect the city from bad weather?

The cult of Priapos seems, as we have seen, mainly to have dealt with sexuality and the phallic power of life with regard to gardening and its protection. The function of Priapos as a weather god was obviously embedded in this hortonomic context, and one might wonder if bad weather and the thieves were understood in analogy. Theft of fruit and plants on the one hand and bad weather on the other threaten the garden's beauty and usefulness in similar ways.[29] Goethe observed in his Roman *Elegies* that protection against thieves is Priapos' main function, but missed the weather aspect,[30] and he furthermore used the god to signify that the artist had been granted fabulous sexual and poetic potency.

The main task of the god, who spent most of his time sleeping, was to protect human gardens from two evils: robbery and bad weather. The deep connection between agriculture and weather religion that we could detect in the Fiji islands obviously also provided the overarching context in Roman and Greek antiquity. The interrelation of fertility, sexuality, robbery, and weather, however, appears to be subtle in the complex world of religion in antiquity. Divine power over growth, weather, health, and evil were compressed in the shape and symbol of the phallus. Its power could drive and keep away evil, while at the same time serving as a symbol of life and reproduction.[31]

Weather could not threaten and impact on the power of this phallus in the slightest, as Cardinal Pietro Bembo writes in his phallus-glorifying poem "Priapos" in the 15th century.[32] Rather, weather and peace, in the sense of being-left-in-peace by severe weather and thieves, constituted central elements in the sphere of reproducing life, flourishing in gardening as well as in human sexuality and eroticism. Androcentrically, if regarded in the context of antiquity's patriarchate, it was the male phallus that held sway over the reproducing forces of life and the fine weather conditions necessary for them.

This androcentric character was common in antiquity as well as in other spheres of the older history of religion, where the gods who reigned over the storm and thunder in particular were typically male. In Greece, *Zeus*, possibly Priapos' father, was both the king of gods and the mighty storm and sky god. Among the Maya, *Chaac* was a violent rain god who demanded sacrifices. *Hadad* was the Semitic storm and rain god in Mesopotamia, who also protected agriculture, and his absence could cause droughts and death. *Moses* may also have cooperated with him when he saved the people from the pursuing Egyptian army and "drove the sea back with a strong east wind and turned it into dry land" (cf. Exodus 14:21). In the Hebrew Bible, *Hadad* is referred to in connection with Yahweh but is also described as a false god, a description that was taken over in Christianity and Islam where he was located as *Beelzebub* among the demons. Different, but still male, is the king of the gods *Indra* in Hinduism, who also is the god of weather and war. One can perceive him especially in rain and thunderstorms, and as a warrior against destructive weather forces. In Norse religion too, the weather was governed by the male and superior – thunder, lightning, oaks, strength, healing, and more were ascribed to *Thor*. The list could be continued with many more male weather gods.

Alongside Priapos, whose popularity probably had to do with his sexual connotations, one should likewise name in the context of antiquity *Jupiter*, regarded by the Romans as equivalent to Zeus. Jupiter is the god of sky and thunder. And as he is regarded as the chief of the gods, weather was obviously ranked among the most significant of the gods' concerns. Despite his being god of the sky, and brother of Neptune and Pluto, his control over life-enhancing weather nonetheless does not seem to have been a focus of belief in Jupiter. Rather, the accord of Jupiter, Neptune, and Pluto expressed their universal power over the space of life, where the first ruled over the sky, the second over the water, and the third over the underworld. Jupiter's power over the sky was deeply connected to the execution of power over the state and social order, the universe, humans, and warfare. One is tempted to say that not only was power in itself regarded as a tool in the hands of men but also the sky and weather in general were ontologically represented as male forces. Could one talk about a cultural code where weather itself was gendered? If there is some truth in such a view, one may wonder if and how this still might influence contemporary worldviews. Are there still traces in modern culture of regarding sky and weather as male, or has weather instead turned into a non-gendered phenomenon?

Healing twins caring for good weather

Not all spiritual forces or deities with power in and over weather lands have, however, been imagined as male in the history of religion, as our third journey will show. This does not lead to a specific region or historical period but to an interesting belief that seems to be translocal and transcultural: the religious significance of twins. From Hinduism in India, through native religions among North American and South American Indians and ancient Maya, to Egypt, to antiquity, and then on to Christian hagiography, one can trace the belief that twins possess a specific spiritual significance and power that, among other things, is also deeply connected to the weather. There is a widespread belief that twins possess spiritual power over nature, and especially over rain and weather.[33]

Scholar of medicine Leon D. Hankoff has explored this significance in detail and traced it through different times and cultures. Belief in the health-giving powers of twins is, according to many ethnographic sources, widespread in mythology, folklore, and religion.[34] In addition to healing, twins are often empowered with the ability to increase fertility, affect the weather, and predict the future. The birth of twins has often been associated with divine power, where the mother has been visited or affected by spiritual beings. According to Hankoff, it was often thought that twins were the result of a superfetation, a kind of double impregnation, one by the man and another by the divine. The power of twins with regard to fertility was in this way anchored in the mode of fertility that allowed their specific coming-into-existence. While Hankoff, in his review and discussion of a broad range of sources, is mostly interested in the connection between mythology and the history of medicine, I will focus here on observations on the weathering power of the healing twins.

Hankoff rightly emphasises the twins as a symbolic representation of the tension and forces in humankind and nature, where they serve as an expression of dualities and as a metaphor for balancing and competing forces within duality.[35] Throughout the history and geography of the planet, twins seem to have been ascribed specific powers, and because we have already observed the interconnection of weather and fertility in agriculture and human reproduction, it should come as no surprise that the belief in twin-power is also connected to the variations of weather.

Twin children among the Tsimshian Indians in British Columbia are reported by Frazer to be able to pray to wind and rain, and in their childhood they can summon any wind by motions of their hands. They can make fair or foul weather and cause rain to fall by painting their faces black and then washing them, representing the rain dripping from dark clouds.[36] Frazer quotes a prayer, "Calm down, breath of the twins", and also emphasises in this context the existence of what he calls "weather-doctors" who can be consulted with regard to requests for specific weather, such as "prop up the clouds that may be lowering" during a celebration.[37] Frazer also notes the way in which bodies are dug up and prepared in a particular way in order to affect the weather, as well as the significance of animals.[38]

Twins can appear among humans and also among gods. The power of Castor and Pollux, for example, is widely depicted in classical texts and images. With regard to weather, they act as saviours of shipwrecked sailors, and senders of favourable winds, besides their capacity to ensure hospitality, to invent music and dancing, and to heal and aid in childbirth.[39] Among African tribes, the mortal twin was (and still is sometimes) believed to have specific skills with regard to the weather.[40] The twin here acted as a life-enhancing force bringing about good weather and abundant rainfall that was good for the crops.

> The twin was considered able to manipulate the forces which made for good
> weather or act as an intermediary for the tribe in obtaining good weather.[41]

Christian believers also received the tradition of the healing twins, as can be clearly traced in the stories of Kosmas and Damian, who were regarded and revered as saints, though mostly because of their miracles of healing rather than because of weather management.

Evidence for twins' early power can be found even in the Bronze Age; for example, archaeologists in the Nordic regions have identified a Bronze Age twin cult that was obviously connected to the sun cult. The figure of the twins in Nordic Fogdarp (Fig. 3.4) served as a decoration for the leaders' horses in the context where twins either symbolically or in person achieved central positions in the power negotiations of the groups.[42] Twins and the sun goddess in relation to each

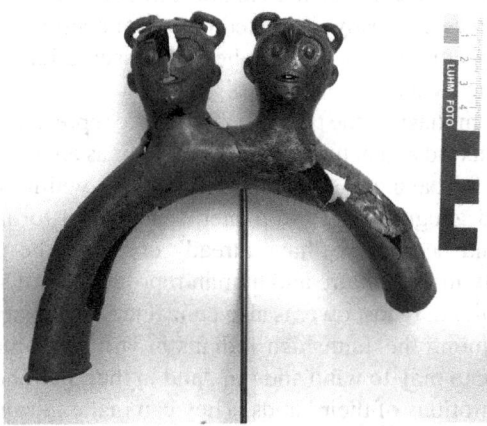

Figure 3.4 Tube with Twins from the Fogdarp Hoard, Bronze Age.

Photo © The Historical Museum at Lund University, by courtesy of the Museum other obviously played an important role in the rituals and symbolic universe of the Scandinavian wetlands of the early Bronze Age.

An enlightening example of the reconciling power of twins, with cosmic relevance for the universe, is found in the Hymn to Asvins in the Indian Rig Veda:

> Like two hands be investing us with vigour,
> like heaven and earth subduing the atmosphere,
> Our songs, Asvins, that proceed towards you,
> sharpen them well, like an axe upon the whetstone.[43]

Another interesting analogy with regard to twins' extraordinary power and their mother's partly divine impregnation is discussed by Hankoff in the context of some Togo tribes that have been analysed by Levi-Strauss. According to Levi-Strauss, using an animal name for a human provides a symbol for relationship between the forces of nature and the human.[44] Twins have sometimes been provided with animal names. Strikingly, the extraordinary powers to influence health, fertility, and weather fall to twins as the inheritance from their unique parentage.

> Closely akin to powers over nature's generosity is the ability to predict the outcome of crops, the weather upon which crops depend, and the offspring of domesticated animals.[45]

Due to his/her parentage, the twin can act as mediator and reconciler between the forces of nature and culture, to which weather also belongs.

The relation between the cosmos and the nature of the twin appears more strongly in ancient Egypt, where their placenta was ascribed a specific significance and identified with the moon. *Khons*, the placenta-moon god, was adored due to his healing abilities. This is only one of many different examples of the complex role of twins in Egyptian religion, where the twin acted as a bridge to the spiritual world.

> Thus, pharaoh and placenta are the material reflections of sun and moon and a basis for an Egyptian religious dualistic formulation.[46]

Due to their specific ability to heal and nurture fertility, the power of the twins to control weather seems to remain limited to good weather, in the sense of ensuring weather that could lead to a good harvest from both plant and animal sources. Twins' power of influencing weather was anchored in their partly divine heritage, and it was deeply interconnected with their skill of healing, a belief that is still alive and well in modern spirituality.

Birds as spiritual messengers – caring for health, fertility, and rain

It is not only humans and gods who perform specific functions in spiritual weather lands; animals can also play a key role. In a new journey, we will travel to the southwest of Native American lands and the ancestral societies of New Mexico, lifting up our eyes to the birds. The Puebloan culture is known for its specific architecture where stone and earth dwellings were built along cliff walls, and where villages (pueblos) were planned as structured community spaces. A road system served religious purposes, as a route for pilgrimage and procession. Furthermore, these pueblos are known for their richly adorned pottery. For rituals, the Puebloans constructed ceremonial structures, known as "kivas" (Fig. 3.5), where a rectangular or round form – earlier one kiva for each 5–6 rooms, later for up to 60–90 rooms in a pueblo – was dug into the earth and entered from the top by ladder.

Weather control and rainmaking seems to have characterised the western Pueblos,[47] while eastern groups seem to have focused more on the healing arts. Religion and life were inseparable in traditional Puebloan culture. The sun represented the Creator. Sacred mountains defined the space and balance of the Pueblo world. Many of the ceremonies revolved around the weather and were devoted to ensuring adequate rainfall,[48] the proper sequence of seasons, abundant rain, and good conditions for sufficient crops. Ceremonies took place in the *kiva* or in the plaza of the pueblo, and priesthoods, for both men and women, were central at these. Furthermore, *kachina* belief, which will be further discussed later, also characterised Pueblo religion. *Kachinas* were ancestors and benevolent spirits that could be symbolically represented in animal, human, or plant forms. *Kachina* dances were ritually performed in order to bring rain and cure illness.[49] Moreover, the conjunction of sun and moon was ritually important for the Pueblo people; the moon was

Figure 3.5 Kiva at Bandelier National Monument, near Santa Fe, New Mexico.
https://commons.wikimedia.org/wiki/File:Bandelier_Kiva.jpg, accessed 26 August 2019

watched carefully for weather omens, for example variously shaded rings that corresponded to different gradations of cold weather.[50]

Aby Warburg, in his study of Puebloan culture, reports a serpent that serves as a weather fetish, and he reproduces a drawing of the serpent and a "worldhouse" that was produced for him by Cleo Jurino.[51]

Studying Puebloan pottery one can find an overwhelming number of designs that have to do with weather, such as clouds of different kinds, rain, snow, wind, lightning, and flowers that are believed to come out of the rain.[52] Many of these pottery designs belong to ceremonies aimed at controlling the weather. Furthermore, the establishment of a new iconic system that archaeologists can identify in Pueblo pottery design of the 14th century was associated with ritual systems regarding fertility and weather. These rituals, in which the *katsinas* also played a key role, probably had the function of integrating disparate social groups in the region. Rituals could therefore be performed in large plazas, visual imagery on artefacts could cut across kin-based group identity and make everyone a member, and the weather rituals served all members of the community and their human ecology.[53] Amazingly, birds seem to have acted as key agents in this interspace between sky and Earth, between the community and its weathered environment. After 1325, birds seem to have become particularly prominent in the Pueblo *kiva* murals.[54]

Following an exciting process of archaeological and anthropological reconstruction of what happened in the American Southwest in Central New Mexico in the 14th century, Suzanne L. Eckert and Tiffany Clark take as their starting point the cultural change with the development of several new ritual systems and religious belief systems in the Pueblo areas in New Mexico between 1300 and 1500. At the core of this process was "the adoption of a complex of icons focusing on fertility, weather control, and community well being".[55]

We can also recognise here, as aforementioned in other fields, the embedding of weather in the matrix of fertility, agriculture, and community well-being. Surprisingly, birds seem to have played a key role in the materialisation of the religious and cultural change. Eckert and Clark studied ceramic, architectural, and faunic data from a selected region and two excavation sites, and they were able to observe how both the imagery and use of birds in ritual contexts increased substantially in two villages during the 14th century. According to their study, this increase corresponded to the development of a new religious ideology that served to integrate diverse populations in the area, obviously in a context of increasing immigration from other areas and a mix of strangers and natives.

Birds are an integral part of modern Pueblo cosmology, mythology, and ritual life (Fig. 3.6). Feathers are used in clothing, hair ornaments, masks, fetishes, and prayer sticks. In earlier times, they served a variety of ceremonial functions. Birds were placed on altars; feathers were used for prayer sticks and as burial offerings. Images of birds were painted on pottery and carved in stone. Hummingbird, eagle, and crow appear in narratives, myths, and songs. Birds can be traced throughout a long history. Among the ritually most important birds were raptors (such as eagles, hawks, falcons, and owls), perching birds, and waterfowl, as well as exotic birds such as parrots and macaws.

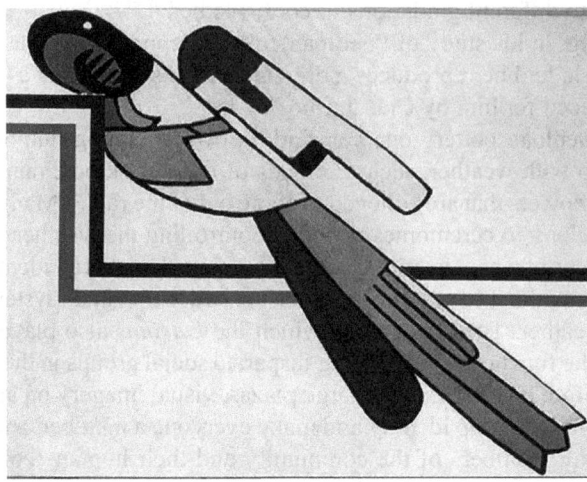

Figure 3.6 Climbing Thick-Billed Parrot, from Kiva 7, layer 31, west wall (after Hibben 1975), in Eckert and Clark (2009), 15

According to Eckert and Clark, birds, and especially their feathers, were used in a wide range of ritual realms vital to community well-being. These included rainmaking, salt procurement, curing, hunting, leadership, and war. Obviously, the increase in using birds in rituals had to do with the need for a new challenging negotiation in times of immigration and the process of integration of multiple social groups.[56] The new ritual regime with birds at the centre emerged as a specific way of facilitating this integration. Rituals were conducted to bring about a change in weather that could serve the growing community's life and well-being, and birds seem to have played a key role in these.

The mural artists who painted birds were obviously familiar with the animals' living actions, as they portrayed them performing specific actions but also depicted them as anthropomorphised figures. *Kiva* murals like this seem to have played an important role in rituals of the 14th-century Pueblo peoples, and thus also in the protection of their community with regard to weather variation (Fig. 3.7).

One might also assume that the variation in the animals found by the archaeologists on different objects indicates some kind of ritual variation where particular birds might have been used in ceremonies following a ritual cycle, even if evidence cannot easily be provided from archaeological traces.[57] One can therefore only speculate as to whether such ritual cycles might have been related to seasonal and weather cycles as well as to the weather-related and environmental behaviours of the living birds.

Evidence can, however, be offered of the substantial increase of birds and feathers in ritual and cultural systems of thought at that time, an observation also true for other parts of the country. The archaeologists can further show that birds were

Figure 3.7 Detail of a Kiva Mural, Kuaua Pueblo, Coronado State Monument, New Mexico, late 15th to early 16th century.

https://commons.wikimedia.org/wiki/File:Kuaua_Kiva_mural.jpg, accessed 2 January 2020

incorporated into a broad variety of rituals among the Pueblo people, and that they played a key role in the human ritual approach to weather and fertility, for the prosperity of the community and the integration of natives and immigrants. Following a quote from Aristotle that a swallow does not make a summer, which reading between the lines also implies that several swallows do in fact indicate summer's arrival, it should not surprise us that birds, as beings at home in the sky and moving continuously between land and sky, are spiritual and physical carriers of meaning and producers of spiritual insights with regard to weather. In brief, birds are integral parts of the flux of weather.

Eckert and Clark indicate further that the increase of birds in Pueblo rituals at that dramatic time was also connected to the emergence of *katsina* religion,[58] a tradition of the *Hopi* people in the same area, in which the *katsinas* represented ancestors who returned as benevolent spirits. These could come down from the mountains as clouds, or up from the earth. They could be male or female and appear as plants, animals, insects, human qualities, the creative force of the sun, and even death (Fig. 3.8).

Katsinas acted as messengers who accepted the humans' gifts and prayers for health, fertility, and rain and carried these back to the gods. Their role as rainmakers was particularly important to the *Hopi*, whose agriculture has always been precarious in the highly arid deserts.[59]

The *Hopi* were certainly aware of this deep relationship between the weather and the spirits, which is emphasised in a traditional story about the *katsinas'* role

Figure 3.8 Drawing of Kachina Doll, in Jesse Walter Fewkes, Dolls of the Tusayan Indians, Leiden: Brill 1894, plate 11, extract.

https://upload.wikimedia.org/wikipedia/commons/8/81/Kachina_dolls.jpg, accessed 26 August 2019

as travelling messengers. Before the *katsinas* depart back to the spirit world, their *katsina* father addresses them in a speech:

> When you go home and get to your parents and sisters and the rest of your relatives who are waiting for you, tell them that they should not wait, but let them come at once and bring rain to our fields. We may have just a few crops in our fields, but when you bring the rain they will grow up and become strong. Then if you will bring some more rain on them we will have more corn, and more beans, and more watermelons, and all the rest of our crops. When harvest time comes we will have plenty of food for the whole winter. So now, this will be all. Now go back home happily, but do not forget us. Come to visit us as rain. This is all.[60]

Might the birds from the Pueblo rituals in New Mexico have inspired the *Hopi katsina* in Arizona?[61] How were the birds in the sky and the Pueblo culture connected to the spirits coming down in the clouds and up from the earth? Obviously the use of feathers and birds also played an important role in *katsina* rituals. Both birds and *katsina* spirits, with their feathers, clearly acted as mediators between humans and the forces of weather variation, mainly for the benefit of the human community and its need for predictable weather, and especially precipitation patterns.

Where are the *katsinas* now, in late modern times? Do they still, in times of global warming and droughts, "visit us as rain"? And can birds, like once among the Pueblos, assist us global nomads to establish peaceful communities of locals and strangers in increasing inter- and transcultural encounters, and comforting weather? Which birds should we engage with, where, and how might they help?[62]

The weather god Yahweh and the Dutch struggle with the sea

Our walk through different cultural landscapes might awaken the impression that the connection between weather, ritual, and belief is reserved for indigenous and premodern contexts, and that religious attitudes and practices with regard to weather variation belong either to small-scale communities, such as the Fijians or Puebloans, or to larger classical civilisations as with Priapos and Jupiter in antiquity. An interesting discussion in the field of Semitic religion, however, shows that there might be some indication that weather is also at work in the roots of belief which flows into the monotheistic, or better Abrahamic, religion of the so-called axial age. Might faith in Yahweh, the one God of the Jewish people, be in relation to or even emerge from belief in the Canaanite weather god? Was Yahweh possibly a weather god in the beginning?

According to scholars in archaeology and biblical studies, the origin of Yahweh, or rather, the expressions of belief in Yahweh, lies in the south of the region. Both archaeological data and biblical texts substantiate the origin of belief in the solitary weather god Yahweh in the late Bronze Age.[63] In an extensive study, Reinhard Müller has shown that Yahweh, the God of the Hebrew Bible, was originally a weather god similar to the Syrian god Baal.[64] His analysis of several texts in the Book of Psalms shows how old cultural songs have been preserved which are from the beginnings of the adoration of Yahweh. In the oldest of these, Yahweh is praised as a mighty fighter who appears in the thunderstorm, defeats his mythical enemies, and bestows rain and fertility on the earth. In younger parts of the historical layer of the texts, Yahweh is praised as the king of gods who, in his reign, protects the order of the universe from the powers of chaos. According to Müller, a comparison with Syrian and Mesopotamian traditions of belief in weather deities makes it evident that the belief in Yahweh did not differ substantially from these. Clearly, not until after the end of the monarchy did the emerging Jewish faith community deeply reinterpret the old hymns to the kingly weather god.

A detailed, rich, and thought-provoking study is offered by Steve A. Wiggins, who explores the meteorological language of the Psalms in the light of our current interest in the weather. "To understand the weather is somehow to glimpse the divine", he states aptly,[65] and investigates in detail the Psalms' biblical sources and their narratives about thunderstorm, wind, rain and clouds, hail, snow, frost, rime, dew, and temperature. Wiggins' study on the Psalms, characterised as a "meteoro-theological survey", confirms even more deeply the other scholars' hypothesis of an early weather god. Wiggins is clear about the very close relationship between

Yahweh and the weather, or should one say a close ontological entanglement of weather and the divine. Weather touches "the very heart of ancient perceptions of the divine".[66] Summarising his observations on the Psalms, Wiggins makes us aware how weather reveals and conceals God, how Yahweh utilises weather as weaponry, and how weather serves to bring glory to the Creator.[67]

Not only the book of Psalms but also other parts of the Hebrew Bible describe the power of God with regard to the forces of weather. As early as the captivity of the people in Egypt, Moses confronts the magicians of Pharaoh when Yahweh causes hail to fall on the Egyptians but not on the Israelites (Exodus 9–10). Further, it was Yahweh who caused a wind to bring locusts and who blotted out the light of the sun for three days. God spoke to the Hebrews and acted through thunder and lightning. On the one hand, the Psalms praise him for sending rain, snow, hail, frost, wind, and sunshine at the appropriate times. On the other hand, one often gets the impression that "Yahweh has all the unpredictability of a god of the storm".[68]

The old weather god Yahweh's power of impacting on the flux of weather was transmitted from the Israelites to the Christians, where Jesus also had the power to command the weather, by calming winds and a stormy sea (Matthew 8:23–27, Mark 4:35–41, Luke 8:22–25) and even walking upon the water of the sea (Mark 6:45–52). Later theologians and saints also continued the tradition of God's power over weather; Gregory of Nazianzus in late antiquity, for example, speaks of God the Creator communicating through creation and responding to social injustice by sending a hailstorm.[69]

Christian saints could also share in this divine power in specific ways.[70] Saint Erosia, for example, was a French martyr who was killed in 714 by the Saracens when she refused to marry a Saracen; she is believed to have the capacity to protect against bad weather, a skill that is also attributed to Medard, bishop of Vemandois in the 6th century. St. Florian, an officer in the Roman army, who was martyred in the times of Diocletian, is believed to be a powerful protector against danger from water and even fire, while St. Genevieve, born in the 5th century, protects against excessive rain, and St. Christopher, the patron saint of travellers, protects against lightning and also protects sailors against storms.

In addition to the old storm god's power, the Virgin Mary became protector of sailors. Serinity Young tells an impressive story about Mary's specific skill in handling the weather. When Mary appeared in Rome after the victory of the Christians over the Turks at Lepanto in 1517 requesting a church to be built, she caused snow to fall just on the spot where she wanted the church and nowhere else.[71] Muslim saints are also believed to partake in the management of the weather, and the saints are believed to be so close to Allah that they have access to his mercy and are therefore able to bring rain. In Turkey and Iran, this belief is evoked in the word for rain, *rahment*, meaning mercy.[72]

Of special interest here is the fact that the old tradition of belief in Yahweh as weather god can continue up to our own days and can influence the construction of national identity, impacting on the politics of national planning, environmental politics, and climate protection. In an illuminating analysis, historian Simon

Schama makes us aware of the history of the Netherlands and the central role that weather and its religiously embedded control, especially of the sea, has played in the process of developing Dutch culture in the golden age and the building of a nation.[73]

Due to its geographical location, the Netherlands is strongly exposed to the forces of the sea and accompanying weather variation, such as winds, storm floods, and rain. Due to its religious history, Christianity, both Catholicism and significantly Calvinist Protestantism, has played a key role in interpreting both the natural environment and the identity of the people and its leaders. Sea weather and faith in God seem to have been intimately related. In a national epic by Jacob Duym from 1600, for example, "sea, wind and polders had fought on the side of the righteous".[74] Analogies are many between the stories in the Hebrew Bible about God's mighty acts relating to weather and the struggle of the Dutch to develop their country in the face of the challenge of rough sea weather. Biblical worldviews of course also catalysed the development of Dutch colonialism. Moreover, the development of the economy was mirrored in the matrix of weather, sea, and faith, when, for example, sinking ships in storms could be interpreted in the light of greed, sin, and disaster.[75] The reclamation of land for humans from the sea in particular was interpreted in a theological lens and loaded with biblical significance. Schama quotes the 16th-century hydraulic engineer Andries Vierlingh: "The making of new land belongs to God alone . . . for he gives to some people the wit and the strength to do it".[76] God has, in his other words, delegated to the Dutch "a kind of license in the act of territorial creation".[77] Settlers in general and the Dutch conquerors of the sea in particular saw themselves as the new children of Israel. The power of Yahweh the weather God, one might say, was transferred to the chosen people, and wind and weather flux was managed by God through them.

Schama further reminds his readers about the synchronicity of the war for national independence and the struggle against the sea.[78] The analogy between the biblical story of the drowning of Egypt's hosts in the Red Sea and the history of the Dutch was particularly significant. The covenant between God and his chosen people continues in the Dutch conquering of the sea and the management of the weather forces that impact on their harsh lands.

> The Almighty had endowed them with the wit and the will to conquer the waters, and even turn the waters against their enemies; and He had raised them to great riches and power, the better to proclaim His omnipotence (rather than to liberate them from it in this earthly world).[79]

Art history and many nationally significant paintings also reveal this pattern. Cornelis van Haarlem, for example, depicts in his *Israelites Crossing the Red Sea* (Fig. 3.9) the people in a common Mannerist style of his time; the historical significance of the Exodus of the Israelites as a patriotic grounding of the Dutch conquering of weather, wind, and sea, as well as of their enemies in political power struggles, is obvious.

Figure 3.9 Cornelis van Haarlem, *Israelites Crossing the Red Sea*, 1594, oil on wood
panel, 42 × 103 cm, Princeton University Art Museum.

https://wikioo.org/paintings.php?refarticle=8DP5LN&titlepainting=Israelites%20Crossing%20
the%20Red%20Sea&artistname=Cornelis%20Cornelisz%20Van%20Haarlem, accessed 2 January
2020

Works of art from the same time also show that the consciousness of the dangers of the sea, especially in storms, was expressed clearly and insightfully and in relation to biblical faith, for example when sinking ships were compared to the chaos of Sodom and Gomorrah, and when the captain of a ship could be compared to Moses, "imploring the Almighty for guidance in leading his flock to safety".[80] Of course, Christ's storm-calming power on the lake of Galilee also offered a prototype for the same pattern, as here in Rembrandt's famous painting (Fig. 3.10).

Pieter Brueghel the Elder's famous painting *The Storm at Sea* (possibly by Joos de Momper rather than Brueghel) (Fig. 3.11) confronts us with a similar line of thought.

On the one hand, we can see here how the painter does justice to the unpredictable and uncontrollable forces of the storm and sea weather; on the other hand, we can sense the churches on the horizon and the light that shines on them, communicating safety and shelter. The barrel and the whale in the centre of the image of ships trying to escape a storm point to the need for the economy to focus on true goods.[81] Rather than the power of the overwhelming sea and weather, it is the force of the faith-supported human sailors to resist this power that seems to be at the core of Momper's painting.

One can only wonder to what extent the old religious code of the active weather god Yahweh in Christian Dutch iconography still lies behind the enormous investments of the Netherlands in dealing with increasing dangerous climate change, and especially the rapid increase in the sea level, via a technologically sophisticated architectural plan for new settlements and modes of urban living directly in and upon the North Sea. Are the Dutch still on their way to conquering the sea, now in climatically changing times, and will they, with God's help, succeed in

Figure 3.10 Rembrandt, *Christ in the Storm on the Lake of Galilee*, 1633, oil on canvas, 160 × 128 cm.

https://commons.wikimedia.org/wiki/File:Rembrandt_Christ_in_the_Storm_on_the_Lake_of_Galilee.jpg, accessed 27 August 2019

winning this struggle against the sea? Is the weather god Yahweh still handing over his power to the chosen people? Or is it rather the image of weather controlling cultural and technical powers that is at the core? Nevertheless, the modern, highly developed society also continues to identify as one that has been selected for those who have wit and will and for those who can mobilise contemporary cultural energy to resist increasing stormy weather and climate-change-caused sea level rise, rather than merely cope and live with it.[82]

In analogy to the Netherlands, the wind also carries a similar central significance in French Provence. Here, the *Mistral* provides a strong force of nature that has powerfully impacted on the environmental and cultural history of the region. Historian Catherine Dunlop is exploring as part of an ongoing project the Mistral's

Figure 3.11 Joos de Momper II (earlier ascribed to Pieter Brueghel the Elder), *Storm at Sea*, between 1610 and 1615, oil on panel, 71 × 97 cm.

https://commons.wikimedia.org/wiki/File:Joos_de_Momper_the_Younger_-_Storm_at_Sea_-_WGA3342.jpg, accessed 27 August 2019

creative influence on painters such as Vincent van Gogh, as well as regional artists.[83] Similarly again to the Netherlands, the violent, cold, and rough northwesterly wind – which in a short time can develop into a threatening storm, and which follows a subtle seasonal alteration – shaped the bodily experiences of the people and blew straight through the interactions of nature and culture. Settlements, houses, and the thatched cottages, which van Gogh was so excited about, were distinctly accommodated to the wind-shaped climate, and the regional inhabitants believed that their strength to withstand was connected to the mythical Mistral. Before the times of industrialisation when steam machines served for energy production, wind power in local mills represented a substantial economic element. Depending on the shape of the landscape, whether it be on the Netherlandish shore or along the Provence valleys, it is exciting to track what the blowing wind, *wetar*, can do to the environment, its culture, history, and self-understanding.

Jaichylyk – the art of weather-making in Kyrgyzstan

As we said earlier, this chapter offers only a few visits to selected places where weather forces and beliefs are interconnected. Many more places could appear on our travelling map such as weather wizards among Polish farmers,[84] thunder magic in Asia,[85] Sami wind deities in the Arctic North,[86] ritual responses to climate change impact among Hindu villagers in the Himalayas,[87] techniques for modern new-spirituality-inspired weather shamanism with teachings directly from the spirits,[88] or the "Hindu declaration on Climate Change" expressing its compassion for the challenges of "population displacement, food and water shortage, catastrophic weather and rampant disease".[89]

Before I try to sum up and reflect on the findings from this chapter, we will set out for a last journey to Kyrgyzstan. Kyrgyzstan is a mountainous country in Central Asia, surrounded by its neighbours Tajikistan, Uzbekistan, Kazakhstan, and China. The Kyrgyz draw on a nomadic past where seasonal migration links sacred places (the *mazars*) with ancestors.[90] Traders and merchants have interacted with the Islamic world since the 10th century and the majority of the Kyrgyz belong to a moderate Islam in which older cultural and shamanistic traditions still play a central role. In particular, regular visits to *mazars* play a key role in the people's spiritual life. In different ways, local popular Islam is entangled with older modes of beliefs and rituals for commemoration, healing, and prayer practices.[91] Kyrgyzstan was a Soviet republic between 1936 and 1991, and in spite of Communist oppression, local Islamic religion was able to survive in the culture's margins, less through doctrine than through its topography of sacred places and its ritual practices. Today, locally based popular Islam again plays a role in the landscape of religion and politics where Islam is promoted as an aspect of national identity, although several external powers from a traditionalist Arabic peninsula and Turkey intervene and compete, and Russian Orthodoxy and Western Protestant mission are also on the stage.[92] An important role in Kyrgyz national identity and also in the believers' spiritual life is held by the old, very long, epic poem *Manas*, which is recited in specific dramatic style by religious specialists at ceremonies. The epic contains copious historical information from throughout the ages, including narratives about warfare, migration and settlement, and traditional livelihoods and lifestyles.

Weather – a phenomenon that is covered by a linguistically interesting rich diversity of words and meanings in the Kyrgyz language – [93] also plays a role in the process of mirroring the people's nomadic past, in preserving and reciting the *Manas* where the Kyrgyz term "jaichylyk" appears. In the context of an ancient military confrontation between the Chinese and the Kyrgyz tribes, the *Manas* tells of the art of weather-making that was performed by the Kyrgyz fighter Almambet in order to overcome the enemies:

> Almambet performed *jaichylyk*.
> And then the highland was covered with hail,
> And the valleys were flooded with rain.
> The clouds dashed across the sky,
> And the thunder roared in the heaven.
> The Chinese who attacked so defiantly
> Were locked up on the ground,
> They dreamed of seeing the sun.[94]

According to Gulnara Aitpaeva, the short description of this weather art, where no details at all are given about its execution, manifests a clear contrast to the much more detailed descriptions of other practices in the *Manas*. Her conclusion is that the epic here presents *jaichylyk* as a secret knowledge and a closed ritual.[95]

Jaichylyk aims at "changing the weather". A *jaichy* is a person with specific skills to influence the weather and to conduct *jaichylyk*. In times of drought, this could involve invoking and bringing rain, and in times of extensive rain that threatened to destroy the hay, it could aim at stopping rainfall. Blizzards could be created in military struggles to conquer an enemy. According to Aitpaeva, who in her insightful study draws on both epic history and oral tradition, the term *jai* contains a diversity of semantic meanings, such as summer, space, time, universe, calmness, order, and spreading. *Jaichylyk* can also mean peace and well-being, and for Aitpaeva, the skill of weather-making is located within a deeper and wider understanding of *jai* as a prosperous world.[96]

In other parts of Central Asia also, weather change is well known as a traditional practice of shamans and other spiritual practitioners. Both in Northern China and in Turkic and Mongolian regions, such practices have characterised popular religions in the spheres of shamanism, animism, and totemism.[97] Among Turkish people and in Kyrgyzstan, spiritual weather mastery was not, however, practiced by shamans or priests but exclusively by spiritually specialised and initiated persons,[98] who were called *Jaichy* in Kyrgyzstan.

The observation that the practice of *jaichylyk* is a secret art conducted by specially skilled persons who are initiated and only known to local people coincides well with both historical and ethnographic knowledge.[99] There is extremely limited information available in the sources, and in the field it is hard to find someone who is willing to talk about it, even if the skill and practice of *jaichylyk* is well known to most of the local people in the country, as I experienced on an extended field trip in 2009. According to the *Manas*, only very few can practise *jaichylyk*, and as a rule these have been men.

Another text in the *Manas*, analysed by Aitpaeva, describes the art of weather-making in the context of a military struggle, where some more details are revealed and where the scholar can identify a conceptual structure of four steps in the *Jaichy*'s act: (a) entering the place and becoming one with the environment (here a lake); (b) achieving harmony with the history and prevalent force of the place (the sun); (c) turning to the spirits of nature (clouds and lands); and (d) using a natural object (a stone).[100] Here we can see how weather-making is not just a simple practice to achieve the intended weather but one connected to a larger cosmic worldview where time and space are believed to be filled with spirits and where the initiated weather maker must have specific secret skills at his disposal in order to communicate with the place, its history, and the spirits. A specific skill to work with chosen objects such as ritual stones is also demanded.[101] Another tool in the practice is the use of verbal formulae which are used in calling on the forces of nature.[102] Not only in military situations but also in disasters the *Jaichy* acts as mediator between the forces of nature and the human population.

Tracing the oral history of contemporary Kyrgyz people, Aitpaeva considers *jaichylyk* rather as a part of people's memories than as an active practice.[103] People remember the existence of a *jaichy* in their village but have no expert among them today. They remember such *jaichys* who could turn a summer day into a winter day, stop rivers, move mountains apart, and call upon the rain. The art of weather-making seems to be closely connected to the skill of experts reciting the *Manas* epic. A deep connection between the experts in reciting, the Islamic spiritual leaders, and the *jaichylyk* can also be identified in the early 19th century. Some evidence is found that *jaichylyk* started to disappear around the middle of the 19th century.[104]

It is difficult to trace the history of weather-making in the context of the contemporary national chaos in the country. The skills of knowing and reciting the *Manas* are still greatly appreciated among the people, and they seem to have rather increased their significance in a situation where ethnic traditions are able to provide some kind of safety and hope in comparison with political corruption and external interventions from other countries, mainly Russia, Turkey, China, and Saudi Arabia. The political strength of Kyrgyz popular Islam with its roots in local ancient religion is without doubt an important force in finding a future path for the country. Ritual activities at sacred places, *mazars*, also seem to play an ongoing central role in the daily life of Kyrgyz people, as I was able to observe in many places with my own eyes (Fig. 3.12).

The challenges of environmental degradation and increasing climate change – which affect the country and human ecology mostly by impacting on long-term weather patterns of existential necessity for farming in vulnerable regions – increase the need to reconstruct spiritual practices promoting harmony between humans and the forces of nature. While every village previously (before the Soviet Union era) had a *jaichy* who could mediate between nature and society, none is left today. Nevertheless, even today one can find those who continue on the path of the old *jaichy*, as Aitpaeva carefully shows. Her analysis keeps our perspective open:

> However, a process of restoring ancient traditional practices in Kyrgyzstan, including *jaichylyk*, is currently underway. Considering that Kyrgyzstan is an economically undeveloped, post-Soviet country, the process is taking different shapes and forms.[105]

The wish to affect the weather, including as a political force, is without doubt strongly alive in Kyrgyzstan today, even if the old ritual and verbal skills might have been lost; this is obvious from a story about how two local people and spiritual practitioners in 2004 hindered the president, who at that time contended illegally for his third term as a leader, from approaching the sacred site and *Manas* mausoleum of his own home region in Talas by "making a wish".[106] The prayer was followed by a rapid weather change where clear weather turned into clouds and heavy snow. The road had to be closed and the president had to cancel his visit.

Figure 3.12 Holy Well and Holy Tree, Mazar "Manjyly Ata", at the southern beach of the
        Yssyk-Kol (holy) lake in Kyrgyzstan.

Photo © Sigurd Bergmann, June 2009. https://seeingthewoods.org/2013/05/31/photo-of-the-week-
sigurd-bergmann, accessed 28 August 2019

The participants interpreted this event as the wish having been accepted, even if
not in the same rich tradition of the old *jaichylyk*.

Whether there really is an ongoing process of *jaichylyk* restoration in Kyrgyzstan or not is impossible to decide from my writing desk in the European
North. Nevertheless, it is obvious that the tradition and remembrance, and maybe
also the reconstruction, of the old Kyrgyz art of weather-making is still nurtured in
the fertile soil of Kyrgyzstan, a country in the midst of tension between global and
local challenges. With regard to global environmental and climatic change, where
particularly the change in water systems has led to a dramatic situation,[107] it is not
easy to strive for a national identity in a complex post-Soviet situation within a
postmodern political world system. Setting one's hope on older ethnic spiritual
traditions that have been historically proven to support the people's safety and

survival in dangerous times appears distinctly meaningful in such a situation, and we wait with bated breath to see if the Kyrgyz people and the new *jaichys* will develop new methods of responding to weather variation challenges, for their own and maybe also for our sake.

One benefit is, as Aitpaeva and other committed scholars in the country empha-sise, the significance of local spirituality as a cultural and political force for emancipation, in a context where religious freedom, but also lived faith, plays an important role for the nation.[108] Without doubt, the art of weather-making, as part of this spirituality, offers exciting traditions, practices, stories, and worldviews for our consideration.

Another is to ask what significance traditional *jaichylyk* might come to have in the late modern context of climate change. One important observation here might relate to the secretness and exclusivity of Kyrgyz *jaichylyk*, which has been strictly reserved for experts, who are obviously not among the publicly known shamans or popular Islamic leaders. Weather-making clearly represented a secret skill that could and should only be administrated and conducted by especially skilled persons who kept their knowledge completely secret. The art of weather-making was regarded as too dangerous for anyone other than a few initiated per-sons, and therefore could not be shared widely.

In contrast to this, modern geoengineering represents a publicly wide-open praxis in science and technology, where the general public seems to be entirely unaware of the dangerous and ethically ambiguous consequences. A national leg-islation system to avoid the most dangerous effects is totally lacking. Hindrances to establishing a critical discourse that might lead to such a regulation are raised by powerful forces with large capital investment interests. As usual in technol-ogy development, the tough competitive processes of innovating, including their expected success and income, are taking place relentlessly and beyond all kind of moral responsibility. The fundamental hypothesis about geoengineering's capac-ity to solve climate change-related problems has never been proven, and it cannot, even as a hypothesis, firmly convince anyone other than those who fanatically believe and invest in it.[109]

Should humans keep their skills of impacting on weather strictly reserved to a few specific situations where the survival of a group is threatened? Or should we accept that every world citizen, and in particular the citizens of the rich nations, is allowed to impact anthropogenically on the weather, globally and locally? Or do we even want to strive for military use of weather-making technology, which is not just fiction but already on its way among the super powers China, the United States, and Russia?[110] How do we cope with the growing insight that the anthropo-centric turn means that the art of weather-making, in the form of climate impact, has resulted in a common global activity where every world citizen through his/her patterns of production, consumption, and mobility affects the weather? Or should we hand over the traditional skills of weather-making to some selected practitioners in science and technology and regard these as inheritors and follow-ers of the old *jaichy*?

None of these alternatives seems to me to be really convincing or promising for a path into a sustainable future for all. Nevertheless, the Kyrgyz story about weather-making together with the other insights from this chapter's journeys might throw valuable light on the contrast between modern practices with regard to weather wisdom and our weather-impacting activity. Hopefully the journeys can sharpen and profile the question as much as possible: do we know what we are doing by impacting on the flux of weather, and do we really aim at doing what we do? And from a moral perspective, do we have a right to make weather, or should we seriously negotiate culturally, spiritually, and politically who has the right to do what? Are we at all capable of understanding ourselves as human weather makers? Compared to the Kyrgyz *jaichy*, we appear, in my view, to be primitive, ignorant, stupid, and self-aggrandising beings in our blindness to how we impact on the weather and are impacted by it.

Faith as a skill in creative weather adaptation in life-giving atmospheres

Finally, I would like to conclude with a reflective discussion. Although this chapter makes no comprehensive attempt to map or systematise religious weather, one can detect some glimpses of a pattern. Weather, in our fields and in our locations, has obviously not been regarded as an overwhelming force that causes Kierkegaard's "infinite resignation", and an existential fear and trembling, but has rather been approached pragmatically. As being alive in weather lands means to be a part of the flux of weather, religions – here regarded as cultural systems[111] – have naturally included and integrated their perceptions of and responses to weather in their practices and worldviews. Being alive as a human being implies being part of the weather. Weather therefore also appears as a fundamental part of faith. Lived religion takes place in a mingling of land, sky, wind, and breath. Weather blows, as "wetar", straight through belief, whether it be in the Fijis, in native Pueblos, on the Dutch coast, or in the Kyrgyz mountains, or whether it be in ancient gardens, with holy twins, or in the flight through the Red Sea.

Perceiving weather in a spiritual lens and responding to it with the toolbox of religious practices and ideologies took different paths in our aforementioned examples. Common to all was to regard the flux of weather as an environmental force that is deeply connected to the power of God, the gods, and spirits or other spiritual forces, as a power that transcends the capacity and horizon of the human community. Weather could appear as the screen against which deities act, or as a divine figure or force itself. Human beings understood themselves as creatures dependent on this power, and they needed to find different modes of approaching it. Continuing the reflections in the first chapter and thinking about weather as something that takes place as the "temperament of being" in the flux between Earth and sky,[112] the religion we have seen in our journeys can also be understood as something that takes place in the flux between Earth and sky.

Beyond this, we can observe that religious responses to the flux of weather are deeply anchored in what we might call uncertainty wisdom. As weather in modern times could for the first time be predicted somewhat reliably with the tools of scientific meteorology, weather in former times caused a deeply unpredictable state of being and exposed the human community and its human ecology to a constant challenge of dealing with weather-based contingency. Religious practices and belief systems obviously had central functions here. Lived faith served as a social and cultural tool for coping with the variation of weather, and especially for dealing with its extremes. Weather variation was constantly perceived through the lens of faith.[113]

A further observation in our aforementioned fields is, unsurprisingly, the embedding of religious weather responses in the human ecology. The needs of agriculture, sea-based mobility, and human fertility in particular appear often as the central contexts in which a religious attitude and practice towards the weather has emerged, whether it be too little or too much rain for cultivating the soil or the winds and storms on the sea that either allowed swift passage or threatened death for seafarers. In military contexts too, weather could appear as a decisive force. Battles could be fought in changing weather, and even the fact that skills of weather mastery were used in a military context shows how deeply interconnected human societies were with changing weather. Surprisingly, often skills of healing are also connected to skills of religious weather mastery, as we saw in the stories about the powerful twins.

A difference between various cultural and religious traditions and geographical contexts might become visible with regard to the active or passive role of humans responding to the flux of weather. In some regions, such as in Oceania, power over the sky was left to the gods, while power over the land, and its crops, was managed by humans. The response to weather was limited to prayers and rituals where God and/or the gods were believed to be the central actors with power. In other regions, however, humans played an active role and developed specific skills, which were often reserved for initiated specialists and spiritual leaders, to impact directly on the change of weather.

In my view, one should depict such skills, which the older history of religion has problematically entitled weather "magic", as spiritual weather mastery. There seems to be a clear difference between humans trying to execute such weather mastery actively, as in Central Asia, and limiting their response more passively to prayers and rituals, and even creating meaning through religious artefacts such as the Pueblonian birds. Even though these weather responses obviously represent an activity that aims at influencing the weather, they were addressed to the masters of weather, God, deities, spirits, or other forces, rather than impacting on weather directly with some kind of mandate from the spiritual powers. God, the gods, and spirits alone, never humans, are regarded and adored as the only real masters of weather. This ancient and traditional code has been fundamentally broken in modern times, when humans themselves can identify as so-called geoengineers, impacting on the weather through reason and technology.

Responses to weather could in this traditional regard either take place as publicly open practices for all, even if led by specialists, or as hidden individual practices that are only visible and open to initiated spiritual experts, in some contexts shamans and in others specific weather experts. Both men and women could actively master the weather, even if this skill in several regions such as Central Asia seems to have been reserved for men. Artefacts such as birds or stones could serve as important instruments for mastering weather forces, while words, prayers, and rituals mostly served as central practical modes of responding to weather.

Deities served a central role in some traditions; they were depicted with manifold attributes, bodily expressions, and specific weather skills, and they have been ascribed complex meaning in broader belief systems. In some traditions, those who hold power over weather in the proper sense were only visible to the experts. Weather gods could have distinct, and also anthropomorphic, profiles such as Priapos, Zeus, Jupiter, or Baal; or they could dwell in the hidden lands of faith such as in the sacred landscape of the Kyrgyz mountains. Moreover, historically weather gods could act in many different ways. While some tamed weather for the sake of protection or punishment, such as Yahweh in the people's flight through the Red Sea, others remained passive as recipients of the believers' prayers, and only indirectly could the weather reveal if their prayers were heard or not. While weather-making experts among the humans could be accused, punished, and even killed after a fatal failure, deities were never accused for how they used their power over and through weather.

Extreme weather, such as droughts, heavy rains, cloudbursts, and thunderstorms, or even seasonal disturbances, could lead to disasters for communities and societies and affect their human ecology and subsistence painfully. Surviving in vulnerable environments meant surviving within weather lands, with the weather and never against it. Being aware of the variation of weather, and being able to relate to its hidden spiritual forces or to God or the deities who hold power over these forces, represented an essential life skill. Sustainability meant, then just as now, to live in harmony with the flux of weather.

Spiritually perceiving and relating to weather increased the subjective and cultural sensibility towards it, and thus advanced the social capacity to adapt to weather variation, especially with regard to weather extremes. The spiritual perception of weather, its religious fabrication of meaning, and responding to it in a diversity of practices and worldviews reveals religion as a central cultural skill for creative adaptation to dangerous environments in general and weather change in particular. This competence represents in this sense once more a crucial skill of making-oneself-at-home on Earth in sunny, cloudy, and demanding weather, in comfortable, life-enhancing weather conditions as well as in troublesome, threatening ones.

Weather, so to speak, constantly blows through the skin and body of human society, continuously testing its ability to adapt. Religion works as a force for climatic resilience. Sensitivity to and empathy with the weather emerge as both spiritually and physically meaningful outworkings of a deep cultural integration of weather and faith where respect and reverence, even fear to some degree, represent practically valuable responses. Wisdom about weather variation makes life liveable with regard to both the unpredictability and uncertainty of the weather to come and the cyclic, seasonal, and natural rhythms of the weather flux. The diversity of religious images and responses to weather variation seems to contribute qualitatively to mirroring and tracking the provoking and challenging task of interpreting the flow of weather, which is so crucial for human survival, even today. And the manifold diverse religious responses seem to be able to assist in navigating the delicate balance between total uncertainty and remembrance-based limited knowledge about weather rhythms in space and time. Weather magic, which Frazer so categorically expelled from the sphere of both religion and scientific reason, appears in such a perspective to be a highly relevant, meaningful, and useful cultural and religious skill for adapting to environmental change and its weather flux.

Whether or not one believes in the Kyrgyz rain makers' ability to impact on rain with their stones, or in the Fijian rituals' power to guarantee fine weather for a good harvest, makes no difference. What matters is the general overarching and underlying skill of identifying oneself as a human being alive within the forces of flux in weather lands, and thus of approaching this flux with reverence and wisdom concerning both its rhythms and its contingency. Science and religion can without doubt interact in this sense and both in different ways nurture an attitude that increases people's sensitivity to the complexity and sacredness of the earth's atmosphere, a discussion that we will mine more deeply in Chapter 5.

At the end of our journey through religions of different weather lands, I would like to suggest an analytical approach beyond the ordinary perspectives. These can vary between weather magic before religion and reason, as with Frazer, and some kind of religious functionalism in small-scale and low-tech societies vulnerable to environmental change, as in contemporary theories of religious studies and anthropology. As indicated previously, such an approach departs from the notion of atmosphere.

While atmosphere represents the central term in modern meteorology, which investigates and to some degree explains the varying condition of the atmosphere at particular times and places, atmosphere in the philosophical sense, and especially in phenomenology, allows a specific focus on the *between*. While the meteorologic atmosphere establishes an object that allows empirical investigation and differentiation, the phenomenological atmosphere dissolves the dualism between object and subject. While meteorology makes it possible to see and explore specific conditions of the one single atmosphere at different places and times, in phenomenology, one is allowed to become aware of atmospheres as constant, both stable and dynamic, spheres of encounter and exchange.

An atmosphere is in this sense characterised by the fact that it does not emerge as a consequence of human actions.[114] An atmosphere surrounds something. It is shining from a living creature, a thing, a place, or an artefact. We can experience atmospheres both intuitively and reflectively. Atmospheres emerge, they can endure, and they can disappear in the spaces *between* us and something. Obviously, humans cannot create atmospheres but they can create artefacts which in themselves are capable of producing and mediating atmospheres.

Atmospheres are characterised through their being both human and physical, both subjective and objective. It is no longer the distinction of subject and object but the encounter of both in a common phenomenon which is at the focus of reflection. Men and women perceive themselves in the mirror of their natural surroundings and their artistically and technologically designed surroundings.

The atmosphere emerges in the interspaces between outer human surroundings and inner bodily-spiritual being. It is not at all diffuse or uncertain, shallow or subjective, but it offers a notion, which in an exciting way expresses the interrelatedness of the outer and inner, the bodily and spiritual, the surrounding and the internalised. In cooperation with related terms such as *aura, chord*, and *charisma*, atmosphere has provided, as I have shown earlier,[115] a fruitful perspective on both the philosophy of nature and the ecotheology of the Holy Spirit in space and place.

With regard to weather, this means that we can approach weather not only in the sense of meteorology but also as a sphere of relation, communication, and integration of human beings and non-human elements and forces of nature. Atmospheres arise in the spaces between the object and the subject, and they impact on both. Weather variation impacts on the emergence and continuity of both human-made and natural atmospheres. Weather, one could say, takes place within atmospheres where the human experiences him/herself both as a part of and as a unique subject within his/her surroundings.

Religious images and responses to weather such as those that we have visited in this chapter can be interpreted as atmospheres in weather lands where humans establish a relation between their own being alive and the forces that flow and blow through this life. Weather is in this sense a life-giving atmosphere that humans cannot affect but on which they are totally dependent. Weather might then be circumscribed as the life-giving atmosphere, a statement that one could give meaning to in both a phenomenological and a meteorological sense. Our final chapter will mine more deeply the potential for a synthesis between spiritual, phenomenological, and meteorological atmosphere-thinking.

Atmospheres which emerge in the spaces between nature and culture, humans and spirits, sky and Earth, and Creation and the Creator are in such a view both necessary life conditions at specific times and places, and interspaces of human bodily and spiritually being. While weather regarded scientifically as a meteorological atmosphere evolves entirely independently of human activities, human life through the lens of faith appears as entirely dependent on the God-given flow of weather and its manifold effects on sky, Earth, and bodies.

Weather as atmosphere is one thing; weather *in* the atmospheres surrounding human souls and bodies is another. However, both are connected in and around the bodily being alive of humans and other life forms. No matter how far advanced technology develops and cyberspaces emerge, weather will remain one of the most crucial life forces ever. Weather will continue to offer crucial conditions for bodily well-being, whether one is able to drink clear water from the tap or one is feeling emotionally well due to the sunshine in- or outdoors. Religious images and responses to weather aim at fertilising such an integration of the weather as meteorological atmosphere, on the one hand, and the life-giving qualities of weather that emerge from this atmosphere and are experienced as a bodily and spiritually vitalising atmosphere, on the other.

Active spiritual mastery of weather, as well as sympathetically approaching weather through liturgies and rituals, would then offer a specific kind of what one in meteorology describes as atmospheric inversion. Inversion in the language of meteorology aims at the reversion of the process where warm air usually decreases in temperature due to increasing height. A state of inversion means that a warm layer of air can function as a lock that does not allow cold air to lift and cool the land under a coverage of warm air above.

Religious images and practices could in a similar way be regarded as impacting on weather as an atmospheric inversion does, as they intervene in and invert the usual flow of weather, either forcefully or more gently through prayers and rituals, and, so to speak, invert the usual flow of weather. The overarching aim of weather faith seems to make it possible to establish such a communicative relation and to experiment with different modes of atmospheric inversion. Sometimes it might really lead to a simple inversion, an impact on the flow of weather that is directly linked to religious activities such as in the arts of rain making. Sometimes, probably most often, this relation remains in an open state where weather still carries on along its own lines and, as stated in the introductory chapter, does not bother at all about us humans.

If the atmosphere of weather takes place (in Ingold's sense) as the temperament of being in the flux between Earth and sky, weather religion locates human being-alive directly within this flux and experiments with creative adaptations to it. Departing from the view of atmosphere allows us to locate the understanding of human being-alive in the temperament of being in the weather flux where the human and his/her surroundings melt together. The religious imagination of and response to the flux of weather takes place in the atmosphere where, drawing on Christian theology, the Trinitarian Spirit appears as a Spirit who gives life both to weather lands and to its inhabitants. God takes place in the atmospheres of weather lands, in and through weather as well as within and between human beings. With many good reasons and as an interpretation of experiencing God within environments, the Trinitarian Spirit of the Christian faith is closely linked through imagination to the metaphors of weather. The Spirit, who gives life and brings the life of the world to come, is blowing, greening, heating, smelling, and blessing. Similar analogies and metaphorical convergences are many in the World Wide Web of weather faith, as we have seen.

The chapter title tries to compress them. *Atmospheres agog* are variations of weather in continuous movement. The flux of weather as atmosphere takes place both in memorable and recognisable rhythms and in totally unpredictable forms that humans regard as chaotic. Religion as a kind of atmospheric inversion intervenes in this flux and allows human beings to identify themselves as both bodily partaking in the weather and spiritually establishing a relation to its hidden forces. Religious imagination of and response to atmospheres agog make it possible to let oneself empower one's life through weather and at the same time try to become an active part, in some way, of its life-empowering rhythm and variation.

Faith in and with weather lands allows one to make oneself at home within the life-enhancing, as well as the life-threatening, flux of weather. Such a religious interpretation of weather and related practices offer a vibrant contrast to the weather image of modernity where the technological doability of everything and short-term meteorological forecasts give the impression of almightiness, and where the fatal ignorance of weather and climate that is impacting on human social, technical, and economic activities in the Anthropocene. Such ignorance obscures the much needed wisdom about uncertainty, sensitivity, and compassion for weather variation, and the humble self-understanding of the human as a partaker in the temperament of being in the flux of weather. The religious attitude that has blown like a wind through this chapter thus represents neither a superstitious nor a magical quackish belief to be attacked by all kinds of criticism of religion but offers a highly reasonable option for grounding one's own being-alive in the surroundings and environments that, due to weather variation, uphold the conditions for life for all.

Weather faith and spiritual weather mastery seem in such a view to have much more in common with modern climate impact science, as one should expect. Pleas for justice in sharing the common gifts of this Earth as the Fijian church leaders formulated on our first journey are grounded in such weather faith as part of Christian faith in the world as creation. One does not need to share the religious faith but can support the same attitude and ideology from many different angles. Pleas for climate justice, anchored in deep weather faith, can be rejected as simple political statements, but they can also be regarded as important cultural skills of creative adaptation, nurtured by deep spiritual cultural wisdom: by weather we live in atmospheres agog.

Notes

1 James G. Frazer, *The Golden Bough: A Study in Comparative Religion*, 2 volumes, London and New York 1890; 2nd edition *The Golden Bough: A Study in Magic and Religion*, 3 volumes, London and New York 1900; 3rd edition 12 volumes, London and New York 1911/1915. Here I use the 3rd edition and mainly Frazer's chapter "V. The Magical Control of the Weather," 62–82.

   For an excellently contextualised evaluation of Frazer's book, see Hans G. Kippenberg, *Die Entdeckung der Religionsgeschichte: Religionswissenschaft und Moderne*, München: Beck 1997, 128–132. Engl. ed., *Discovering Religious History in the Modern Age*, Princeton, NJ: Princeton University Press 2002.

2 J.M.W. Turner, *The Golden Bough*, exhibited 1834, oil on canvas, 1041 × 1638 mm, www.tate.org.uk/art/artworks/turner-the-golden-bough-n00371, accessed 21 August 2019. Frazer interprets Turner's painting carefully at the beginning of his book, James G. Frazer, *The Golden Bough: A Study in Comparative Religion*, 2 volumes, London and New York 1890, Vol. 1, 11a. Cf. Jonathan Jones, "Modern Myths," *The Guardian* 10 December 2005, www.theguardian.com/artanddesign/2005/dec/10/art.classics, accessed 21 August 2019.

3 Frazer, op.cit., 3rd edition, *Preface*, London June 1922, 10a.

4 Ludwig Wittgenstein, "Bemerkungen über Frazers *Golden Bough*/Remarks on Frazer's *Golden Bough* written 1931–36, first published in 1967," in: *Synthese*, Vol. 17, Dordrecht Holland: D. Reidel Publishing Co, 1967, 233–253. Quoted from *Ludwig Wittgenstein: Philosophical Occasions, 1912–1951*, ed. by James C. Klagge and Alfred Nordmann, Indianapolis: Hackett Publishing 1993, 118–155, 119.

5 Ádám Molnár, *Weather Magic in Inner Asia*, Bloomington, IN: Indiana University, Research Institute for Inner Asian Studies 1994.

6 The following study is most careful and well thought-through: Serinity Young, "Religion and Weather," in: Stephen H. Schneider, Terry L. Root, and Michael D. Mastrandrea (eds), *Encyclopedia of Climate and Weather, Volume 1 Abs-Ero*, 2nd edition, Oxford: Oxford University Press 2011, 3–8.

7 Simon D. Donner, "Domain of the Gods: An Editorial Essay," *Climatic Change* 85, 2007, 231–236.

8 Ibid., 231.

9 Ibid., 231.

10 Ibid., 233f.

11 For a detailed analysis of "unusual weather" in Fiji as effects of El Niño 1982–1983 and the responses, see 90–101.
Cf. Tim Bayliss-Smith, Richard Bedford, Harold Brookfield, Marc Latham, with contributions from Muriel Brookfield, *Islands, Islanders and the World: The Colonial and Post-colonial Experience of Eastern Fiji*, Cambridge: Cambridge University Press 1988.

12 Simon D. Donner, "Making the Climate a Part of the Human World," *Bulletin of the American Meteorological Society* 92, 10, October 2011, 1297–1302, 1298. http://dx.doi.org/10.1175/2011BAMS3219.1.

13 Thomas Williams, *Fiji and the Fijians*, ed. by George Stringer Rowe, Boston, MA: Congregational Publishing Society 1871, 183 (1st edition, London: Alexander Heylin 1858), https://archive.org/details/fijifijians01will/page/n7, accessed 20 August 2019.

14 www.presbyterian.org.nz/sites/default/files/speaking_out/Moana_Declaration_2009_Pacific_Churches.pdf, accessed 20 August 2019. The declaration was reaffirmed in the "Statement of the Pacific Conference of Churches on Climate Change and Resettlement," 2011, http://climatepasifika.blogspot.com/2011/05/statement-of-pacific-conference-of.html, accessed 20 August 2019.

15 For more detail, see Upolu Lumā Vaai, "'We Are Therefore We Live': Pacific Eco-Relational Spirituality and Changing the Climate Change Story," in: *Toda Peace Institute, Policy Brief No. 56*, October 2019, Tokyo: Toda Peace Institute 2019.

16 John D'Arcy May, *Christus Initiator: Theologie im Pazifik* (Theologie Interkulturell 4), Düsseldorf: Patmos 1990, 42.

17 Ibid., 96f.

18 Ibid., 104.

19 Ibid., 132.

20 Regarding climate change damage to the islands, see S. George Philander (ed.), *Encyclopedia of Global Warming and Climate Change*, 2nd edition, Thousand Oaks, CA: Sage 2012, 555. Regarding climate politics in Fiji, see Randall S. Abate and Elizabeth Ann Kronk (eds.), *Climate Change and Indigenous Peoples: The Search for Legal Remedies*, Cheltenham and Northampton, MA: Edward Elgar Publishing 2013, 375, and the *Nadi Bay Declaration on the Climate Change Crisis in the Pacific*, 31 July 2019, https://cop23.com.fj/nadi-bay-declaration-on-the-climate-change-crisis-in-the-pacific/, accessed 20 August 2019.

21 According to http://thediplomat.com/2016/02/cyclone-winston-wreaks-havoc-on-fiji/, accessed 7 March 2016.

22 Reproduction in Williams, op. cit., 158, https://archive.org/details/fijifijians01will/page/158, accessed 20 August 2019.

23 Cf. UNESCO Pacific Member States, *Traditional Knowledge for Adapting to Climate Change*, Apia, Samoa: UNESCO, and Yuseong-gu, Daejeon, Korea: ICHCAP 2013, 16–17.

24 Williams, op. cit., 73.

25 Cf. Hans Herter, "Priapos (1)," in: *Paulys Realencyclopädie der classischen Altertumswissenschaft (RE) XXII*, Stuttgart: Metzler 1954, 1914–1942; and Hans Herter, *De Priapo*, Giessen: Töpelmann 1932 (reprint: Forgotten Books 2018).

26 Cf. Javier Angulo (a), Pedro A Fernández Vega (b), and Marcos García (c), "Sexualidad y erotismo en el mundo grecorromano / Sexuality and Eroticism in the Greco-Roman World," *Revista Internacional de Andrología* 6, 2, 2008, 140–151.

27 Ingo Vetter, *Priapos trädgård*, Gävle konstcentrum 2009, www.gavlekonstcentrum.se/ images/dokument/priapos_garden.pdf, accessed 8 March 2016.
28 Magnus Jensner, writing about Ingo Vetter's art work and its "arcadian blues", www. gavlekonstcentrum.se/images/dokument/priapos_garden.pdf, accessed 22 August 2019.
29 Song 61 in the *Priapeia* interconnects weather and robbery:

   "Neither the winds nor rains nor yet the droughts
   Caused just complaining to the apple-tree;
   Nor me the starling or the robber Daw
   Or crow as crone old-grown or watery goose
   Or thirsty raven e'er endamagèd.
   No! but from bearing scribblers' rubbish verse
   On labouring branches comes mine every woe".

   PRIAPEIA sive diversorum poetarum in Priapum lusus or SPORTIVE EPIGRAMS ON PRIAPUS, by divers poets in English verse and prose, translation by Leonard C. Smithers and Sir Richard Burton, 1890, www.sacred-texts.com/cla/priap/prp61.htm, accessed 8 March 2016.
30 Johann Wolfgang Goethe, *Erotica Romana*, I, 1788–1790:

   "You, stand here at my side, good Priapus – albeit from thieves I've
   Nothing to fear. Freely pluck, whosoever would eat".

   www.gutenberg.org/files/7889/7889-h/7889-h.htm#link2H_4_0001, accessed 22 August 2019.
31 Cf. Catherine Johns, *Sex or Symbol? Erotic Images of Greece and Rome*, New York: Routledge 1982, 52.
32 ". . . Und er ist

   Immer der gleiche, ob der Himmel glühet im Hundsstern,
   Oder ob glitzernder Reif decket das Wintergefild.
   Nie verwelket er jemals, da ihm das Wetter nichts anhat,
   Und es gibt keinen Ort, wo man vergebens ihn pflanzt".

   Kardinal Pietro Bembo (1470–1547), *Ode Priapus,* übersetzt von Albert Wesselski, zit. n. Eduard Fuchs, *Illustrierte Sittengeschichte*, Band 4, München: Albert Langen, 1909, 264.
33 Frazer made us aware about the twins' power in his Wittgenstein, op. cit., 67a.
34 Leon D. Hankoff, "Why the Healing Gods Are Twins," *The Yale Journal of Biology and Medicine* 50, 1977, 307–319.
35 Ibid., 307.
36 Frazer, op. cit., 67a.
37 Ibid., 68.
38 Ibid., 71f.
39 Ibid., 310.
40 Among some tribes, though, they were killed as threatening agents of evil forces. In some places, this is still the case today.
41 Frazer, op. cit., 311.
42 Cf. Tova Lindblad, *Fogdarpsfyndet: En landskapsanalys av ett depåfynd från den yngre bronsåldern*, kandidatuppsats vid Uppsala Universitet, Institutionen för arke- ologi och antik historia, 2019, www.diva-portal.se/smash/get/diva2:1325446/FULL TEXT01.pdf, accessed 23 August 2019.
43 The hymn *To Asvins*, Rigveda, quoted in Hankoff, op. cit., 312.

44 Hankoff, op. cit., 314.
45 Ibid., 314.
46 Ibid., 316.
47 Trudy Griffin-Pierce, *Native Peoples of the Southwest*, Albuquerque: University of New Mexico Press 2000, 42.
48 Barry Pritzker, *A Native American Encyclopedia: History, Culture, and Peoples*, Oxford: Oxford University Press 2000, 73.
49 Donald Ricky, *Native Peoples A to Z: A Reference Guide to Native Peoples of the Western Hemisphere*, Vol. 8, 2nd edition, Hamburg MI: Native American Books 2009, 2061.
50 J. McKim Malville and Claudia Putnam, *Prehistoric Astronomy in the Southwest*, revised edition, Boulder, CO: Big Earth Publishing 1993, 24.
51 Aby M. Warburg, *Images from the Region of the Pueblo Indians of North America*, Ithaca and London: Cornell University Press 1995, 10f. Warburg has completed the drawing with his own annotations:
    "The serpent (Ttzitz Chu'i) and the cosmological drawing with the weather-fetish were sketched for me on 10 January 1896 in my room, no. 59, in the Palace Hotel in Santa Fe, by Cleo Jurino, the guardian of the Estufa at Cochita. C. J. is also the painter of the wall-paintings in the Estufa. The priest of Chipeo Nanutsch. 1. Aitschin, house of Yaya, the fetish. 2. Kashtiarts, the rainbow. 3. Yerrick, the fetish (or Yaya). 4. Nematje, the white cloud. 5. Neaesh, the raincloud. 6. Kaasch, rain. 7. Purtunschtschj, lightning. 10. Ttzitz-chui, the water-serpent. 11. The 4 rings signify that whoever approaches the serpent and does not tell the truth, drops dead before one can count to 4". *Bilder aus dem Gebiet der Pueblo-Indianer in Nord-Amerika* offers a transcript of a slide lecture by a patient in Ludwig Binswanger's Kreuzlingen Sanatorium on 21 April 1923. Warburg did not regard the text as publishable, and a translation was first published after his death, initially in 1938–1939, and the full text in 1988 (*Serpent Ritual: An Account of a Journey*, Berlin: Wagenbach 1988), www.jstor.org/stable/10.7591/j.ctt1g69xgc.5, accessed 26 August 2019.
52 Ruth Leah Bunzel, *The Pueblo Potter: A Study of Creative Imagination in Primitive Art*, New York: Dover Publications 1972, 70 (1st edition New York: Columbia University Press 1929).
53 Cf. Suzanne L. Eckert, *Pottery and Practice: The Expression of Identity at Pottery Mound and Hummingbird Pueblo*, Albuquerque: University of New Mexico Press 2008, 92.
54 Polly Schaafsma, *Cultural Attitudes to Animals Including Birds, Fish and Invertebrates: Learning from Art about the Cultural Relationships between Humans and Animals*, Department of Archaeology, University of Southampton 1986, 95.
    Schaafsma also suggests that the roadrunner's tracks might have served as records of myths and ceremonies. Polly Schaafsma, "Supper or Symbol: Roadrunner Tracks in Southwest Art and Ritual," in: Howard Morphy (ed.), *Animals into Art*, London and New York: Routledge 2014, 253–267.
55 Suzanne L. Eckert and Tiffany Clark, "The Ritual Importance of Birds in 14th-Century Central New Mexico," *Journal of Ethnobiology* 29, 1, 2009, 8–27, 8.
56 Ibid., 12.
57 Ibid., 23.
58 Ibid., 24.
59 Cf. the online exhibition *Rainmakers from the Gods at the Peabody Museum of Archaeology and Ethnology*, Cambridge, MA: Harvard University Press 2015, www.peabody.harvard.edu/katsina.
60 Ibid., www.peabody.harvard.edu/node/2026.
61 On Jesse Walter Fewkes' suggestion, see W. Jackson Rushing III, *Native American Art in the Twentieth Century: Makers, Meanings, Histories*, London and New York: Routledge 2013, 24.

62 In Northern European contexts also, birds have historically been related to weather observation. Weather proverbs in folk sayings use the weathercock not only on the roof but also for weather prediction. In my childhood, farmers used to recite, not very helpfully: "Wenn der Hahn kräht auf dem Mist ändert sich das Wetter oder es bleibt wie es ist". (If the cock crows on the dung heap, the weather will change or stay as it is.) Low flying swallows indicate rain (quite a certain method as low pressure impacts on the height of flying insects). Bad weather is indicated by birds sitting together on wires, or fish jumping out of the water. Russian country lore mentions sparrows hiding in the bushes and crying bullfinches as indicators of the approaching winter.

63 Martin Leuenberger, "Jhwhs Herkunft aus dem Süden: Archäologische Befunde – biblische Überlieferungen – historische Korrelationen," *Zeitschrift für die Alttestamentliche Wissenschaft* 122, 1, March 2010, 1–19; Cf. Martin Leuenberger, "Noch einmal: Jhwh aus dem Süden. Methodische und religionsgeschichtliche Überlegungen in der jüngsten Debatte," in: M. Meyer-Blanck (ed.), *Gott und Geschichte*, Vol. XV, Leipzig: Europäischer Kongress für Theologie (VWGTh 44) 2016, 267–287.

64 Reinhard Müller, *Jahwe als Wettergott: Studien zur althebräischen Kultlyrik anhand ausgewählter Psalmen*, Berlin: Walter de Gruyter 2008.

65 Steve A. Wiggins, *Weathering the Psalms: A Meteorotheological Survey*, Eugene, OR: Cascade 2014, 3.

66 Ibid., 3.

67 Ibid., 154–156.

68 Serinity Young, "Religion and Weather," in: Stephen H. Schneider (ed.), *Encyclopedia of Climate and Weather*, Oxford: Oxford University Press 1996, 639–643, 639.

69 According to the patristic theologian Gregory of Nazianzus, it is through creation itself that God guides, instructs, and proclaims himself to human beings (*Oratio* 38.13; 16.5; 6.14). Sigurd Bergmann, *Creation Set Free: The Spirit as Liberator of Nature* (Sacra Doctrina: Christian Theology for a Postmodern Age 4), Grand Rapids Michigan and Cambridge, UK: Eerdmans 2005, 108. In his oration on the failed harvest resulting from a hailstorm (*Oratio* 16.18), Gregory maintains that the cause was actually the wealthy's injustice towards the poor; God intends to teach the wealthy a drastic lesson now. See Bergmann, *Creation Set Free*, op. cit., 88.

70 The following examples are described at www.catholic.org/saints.

71 Young, op. cit., 640.

72 Ibid.

73 Simon Schama, *The Embarrassment of Riches: An Interpretation of Dutch Culture in the Golden Age*, Berkeley, Los Angeles and London: University of California Press 1988.

74 Ibid., 27.

75 Ibid., 32.

76 Ibid., 35.

77 Ibid.

78 Ibid., 37.

79 Ibid., 45f.

80 Ibid., 32.

81 This has been interpreted according to a common saying at that time: If the whale plays with the barrel that has been thrown to him and gives the ship time to escape, then he represents the man who misses the true good for the sake of futile trifles.

82 For a national response to climate change and rising sea levels demanding a re-examination of the country's water management, see *Water Management in the Netherlands*, by the Dutch Ministry of Infrastructure and Environment: Rijkswaterstaat 2009, https://web.archive.org/web/20140221225045/www.rijkswaterstaat.nl/en/images/Water%20Management%20in%20the%20Netherlands_tcm224-303503.pdf, accessed 27 August 2019, and for the urgency of the increasing challenge, comparing

Bangladesh and the Netherlands, see *Sea-Level Rise in Bangladesh and the Netherlands: One Phenomenon, Many Consequences*, Bonn: Germanwatch 2004, https://germanwatch.org/sites/germanwatch.org/files/publication/3642.pdf, accessed 27 August 2019; and Peter Kuipers Munneke, "The Question Is Not if the Netherlands Will Disappear Below Sea Level, But When," *Utrecht University News* 8 August 2018, www.uu.nl/en/news/the-question-is-not-if-the-netherlands-will-disappear-below-sea-level-but-when, accessed 27 August 2019.

83 Catherine Dunlop, *Force of Nature: The Mistral in French History*, Munich: Lunchtime Colloquium at the Rachel Carson Institute for Environment and Society, 31 October 2018; Cf. C. Dunlop, "Looking at the Wind: Paintings of the Mistral in Fin-de-Siècle France," *Environmental History* 20, 3, 2015, 505–518.

84 Urszular Lehr, "Weather Wizards and Contemporary Protecting Resources," *Acta Ethnographica Hungarica* 54, 2, 2009, 309–318.

85 Florian C. Reiter, *Basic Conditions of Taoist Thunder Magic*, Wiesbaden: Harrasowitz Verlag 2007.

86 Among the Sami the god of wind, *Biegga-almmái*, ruled the wind and was important particularly because he was able to move reindeer by making the wind blow continuously in one direction. He is generally portrayed with two shovels in his hands, which he used to shovel the winds into and out of his cave.

87 Ehud Halperin, "Winds of Change: Religion and Climate in the Western Himalayas," *JAAR (Journal of the American Academy of Religion)* 85, 1, 2017, 64–111.

88 Nan Moss and David Corbin, *Weather Shamanism: Harmonizing Our Connection with the Elements*, Rochester Vermont: Bear & Co 2008.

89 www.hinduismtoday.com/pdf_downloads/hindu-climate-change-declaration.pdf and www.hinduclimatedeclaration2015.org/english, accessed 30 August 2019.

90 David W. Montgomery, "Kyrgyzstan," in: Mark Juergensmeyer and Wade Clark Roof (eds.), *Encyclopedia of Global Religion*, Thousand Oaks, CA: Sage 2011, 681.

91 Cf. on the Kyrgyz *mazars* as "sacred places of earthed faith" Sigurd Bergmann, "Religion im gefährlichen Umweltwandel: Zur Herausforderung der Beheimatung im Raum," in: Arnd Heling (ed.), *Der Ostseeraum und seine Wälder: Nachhaltigkeit im Zeichen des Klimawandels*, München: Oekom 2010, 232–245.

92 Regarding the situation and development of a Kyrgyz Christianity that challenges the common belief that being a Kyrgyz implies being a Muslim, see David Radford, *Religious Identity and Social Change: Explaining Christian Conversion in a Muslim World*, New York: Routledge 2015.

93 Dinesh Bhugra, Samson Tse, Roger Ng, and Nori Takei (eds.), *Routledge Handbook of Psychiatry in Asia*, Abingdon and New York: Routledge 2016, 15.

94 According to Gulnara Aitpaeva, "Jaichylyk: Harmonizing the Will of Nature and Human Needs," in: Dieter Gerten and Sigurd Bergmann (eds.), *Religion in Environmental and Climate Change: Suffering, Values, Lifestyles*, London and New York: Continuum 2011, 337.

95 Ibid., 338.

96 Ibid., 333–335.

97 *History of Civilizations of Central Asia*, Vol. IV, Part 2, edited by C.E. Bosworth and M.S. Asimov, New Delhi: Motilal Banarsidass Publishers 2003 (UNESCO 2000), 65–68.

Devin DeWeese, *Islamization and Native Religion in the Golden Horde: Baba Tükles and Conversion to Islam in Historical and Epic Tradition*, University Park, PA: Pennsylvania State University Press 1994. Regarding the mastery of weather among the Turks in the 6th and 7th centuries, see p. 176, and on the struggle between Northern nomads and the Chinese that was decided by a weather-control contest with hailstorms between the Buddhist Lama of the nomads and Vankhas and the Arabs who came to rescue China, see p. 178.

98 Emilia Pásztor, "Prehistoric Sky Lore and Spirituality," in: Dragoş Gheorghiu (ed.), *Archaeology Experiences Spirituality?* Cambridge: Cambridge Scholars Publishing 2011, 89–116, 95.

99 Many important details are presented by Molnár, op. cit., 36f, 48f, 50f, 64f, 88f, 92f, and 116f.

100 Aitpaeva, op. cit., 342.

101 The usage of ritual stones in spiritual weather mastery also seems to be common in other parts of Central Asia. Cf. Molnár, op. cit., 75.

102 Aitpaeva, op. cit., 344.

103 Ibid., 346.

104 Cf. for the details Aitpaeva, op. cit., 350–353.

105 Aitpaeva, op. cit., 356f.

106 Ibid., 357f.

107 For a detailed description of how climate change affects the water system of Kyrgyzstan, see Beatrice Mosello, *How to Deal with Climate Change? Institutional Adaptive Capacity as a Means to Promote Sustainable Water Governance*, Heidelberg: Springer 2015, 117–159.

108 Cf. in detail on the relation of religion and nationality Melissa M. Wilcox, *Religion in Today's World: Global Issues, Sociological Perspectives*, New York: Routledge 2013, 154f.

109 Cf. Mark G. Lawrence, Stefan Schäfer, Helene Muri, Vivian Scott, Andreas Oschlies, Naomi E. Vaughan, Olivier Boucher, Hauke Schmidt, Jim Haywood, and Jürgen Scheffran, "Evaluating Climate Geoengineering Proposals in the Context of the Paris Agreement Temperature Goals," *Nature Communications* 9, article number: 3734, 2018; Christopher J. Preston (ed.), *Climate Justice and Geoengineering: Ethics and Policy in the Atmospheric Anthropocene*, London: Rowman and Littlefield 2016; and Forrest Clingerman and Kevin J. O'Brien (eds.), *Theological and Ethical Perspectives on Climate Engineering: Calming the Storm*, Lanham: Lexington Books 2016.

110 To get an impression of the unbelievable risks and dangers of military weather manipulation with meteorological and technological instruments, see Uwe Laub's novel "Sturm" (München: Heyne 2018), which is built on actual information about ongoing experimental activities among the super powers. We will discuss the military use of meteorology further in Chapter 6. Cf. https://uwelaub.de/hintergrundinfos-zu-sturm/, accessed 28 August 2019.

111 In the sense of Clifford Geertz, "Religion as a Cultural System," in: *The Interpretation of Cultures*, New York: Basic Books 1973, 87–125; Cf. Sigurd Bergmann, "Religion, Culture and God's Here and Now: Contextual Theology in Dialogue with Social Anthropology," *Svensk Teologisk Kvartalskrift* 81, 2, 2005, 67–76.

112 Tim Ingold, *Being Alive: Essays on Movement, Knowledge and Description*, Abingdon: Routledge 2011, 130.

113 Learning from Hultkrantz' theory about the human ecology of religion one can also state here that it is not that nature, land, or in our case weather directly impacts on the form of religious imagination and practice, but that the impact is always filtered through the human ecology, namely how human life, subsistence, and survival are affected by different landscapes and varying weather. Cf. Åke Hultkrantz, "Ecology," in: Mircea Eliade (ed.), *The Encyclopedia of Religion*, vol. 4, New York: Macmillan 1987, 581–585, and "An Ecological Approach to Religion," *Ethnos* 31, 1–4, 1966, 131–150.

114 Gernot Böhme, *Atmosphäre: Essays zur neuen Ästhetik*, Frankfurt am Main: Suhrkamp 1995.

115 Sigurd Bergmann, *Religion, Space and the Environment*, London and New York: Routledge 2014, 37–44.

4 Weathering the history of Christianity

Justice, witchery, and moral thunder

The interconnections between weather, religion, and culture are rich and complex, as we saw in the previous chapter. This chapter will focus on the Christian interpretation of weather in different periods of Christianity's history. Interesting here is the fact that weather was primarily perceived as a sphere of God's sovereignty and that human influence on it, through morally dubious activities, was regarded as an act of evil rather than a process of synergy and cooperation between humans and the Divine.

After selected excursions to visit theologians in late antiquity in the East and the medieval scholastic West, as well as to get a glimpse of Shakespeare, the chapter will take us deeper into two selected periods in European history. One deals with the condemnation as witches of certain women who demonstrated the ability to affect the weather and use it for their own destructive purposes. These women were, in the late medieval period, accused and persecuted as so-called *weather witches*. In the 18th century, we will trace another interesting theological approach to weather, where German Protestant preaching interpreted thunderstorms, failed harvests, and droughts as signs from God, as strong calls to repentance.

Weather as a device in the history and drama of salvation

As we saw earlier, the heritage from the Hebrew Bible where Yahweh had at his command the powers of weather-making was also handed down to the Christian faith. If the image of God in ancient Jewish times emerged from the belief in a Canaanite weather deity,[1] it should be no surprise that Jesus in the stories of the New Testament, and the triune God further on in the history of salvation, acts through, and communicates with believers in, the sphere of weather, mainly the atmospheres of the sky. The space of weather change was therefore perceived as a system of signs where one could read and understand the relation of God to his people and to the world in general. Weather became a kind of moral *barometer*.[2] States of "good" weather that enhanced the human ecology of the community could be interpreted as expressions of God's love and care, while states of "bad" weather, which threatened agriculture and survival and damaged the social structure, could be interpreted as expressions of anger and punishment for misbehaviour in the human community.

Similarly to the Greek gods, the Jewish-Christian God also used weather as an expression of power. In Psalm 29, God does not just command weather but seems to be the weather itself: while "the voice of the LORD is over the waters", "the God of glory thunders", and "the voice of the LORD is powerful" and "majestic", it is "the LORD who breaks in pieces the cedars of Lebanon" (NIV).

A further vivid example of how God manages the weather is given by the Eastern theologian Severian in Syria in the 5th century. In his view, God has created the clouds as pipes through which he scoops up the salty water of the sea.[3] God lays his immortal, invisible hand over the clouds so that they do not empty all at once. The whole image reminds us of a pumping station where God's finger decides how much or how little rain is sent as a gift to the earth. According to the 6th-century Eastern Church topographer Cosmas Indikopleustes, the air was originally committed to the devil, but since his fall, the angels have undertaken his duties.[4]

Obviously, weather serves as a symbol and physical expression of God's power in and over creation. This power can be used for vengeance and punishment, as in the story of the flood, but also as a revelation of love and care as in the sign of the rainbow and the covenant between Creator and creation.[5] In Christian belief, deeply rooted as it is in the Jewish tradition, weather becomes primarily a moral force. Weather turns into a dramatic stage where God reveals his relationship with the created. Lifting one's eyes to the sky implies encountering the weathering God in weathered lands. To live in this creation is to live at the mercy of a loving God who expresses his mercy through the flux of wind and weather. In this way, weather change represents a crucial part of the history of salvation.

The category of justice in particular seems historically to have played a key role herein. When threatening weather flowed in, in the form of either thunderstorms and floods or unusual seasonal change, believers interpreted the events on the weather scene in relation to the community of humans. Faced with extreme weather events, the communion was investigated with regard to sin and violations of social justice. Injustice on a larger scale could be punished by the Creator, who sent threatening weather with a pedagogic intention. Humans ought to learn how to mirror God's love to all, and especially to the poor, and demanding weather was a tool by which God could open the eyes and minds of humans, so that they would repent and take another path. Because the environmental variation (and mainly the variation of rain and sunshine) in created weather lands served, in itself, to express the Creator's love to all, human social life had to mirror this. Likewise God's justice ensured that the goods of natural life are distributed to all created beings. The Creator's justice and love towards creation demanded that we pursue this equality of all and a just distribution of God's gifts of life. Threatening, damaging weather could therefore be read as a sign of God's anger but also as pedagogical help from the loving and merciful God who sent bad weather in order to get his people on the right path again. Weather disasters in this way had their roots in the unjust order of social life, rather than in simple natural processes beyond the control of human beings.

A similar moral code seems to have survived up to our own day; now, too, dangerous climatic change is related to the modes of human social production and consumption, certainly with the difference that today these can be traced scientifically, while in former times, only God kept watch over human activities and reacted through altering the weather. While the hypothesis of "god" is nowadays taken out of the equation and humans are able to be aware of how they are directly impacting on the changing climate and weather, the old code about a deep interconnection between threatening atmospheric change and fatally unjust and unsustainable human social behaviour is returning on a global scale, though with the difference that the formerly critical role of the Creator God is now played by both the atmosphere itself and its human interpreters in climate science and climate-committed citizenship. Or perhaps there has rather been a change in roles, where the ancient Creator God has been replaced by other real and/or imagined forces in between humankind and weather, such as "Gaia" or the presumed "Earth system" that reacts and responds to human activities as a kind of subject? Again, the skill of spiritually reading weather and of projecting one's observations, interpretations, and feelings onto the screen of the whole of the earth as one common weather space is tempting under the new conditions of global climate change, and it by no means needs to turn into a speculative subjectivist esoteric escapism. My question is, rather, whether and how modern weather wisdom, enlightened by ancient traditions in human weather history, might develop into a force that can interact with science and committed climate politics. This is a question that we will return to in the book's final chapter. Can it become a "dramatic device" in the forthcoming struggle to cope with dramatically and dangerously changing weather?

Let us, before we return to historical perspectives from late antiquity, cast a quick glance over Shakespeare's early modernity in the 16th and 17th centuries, where weather seems to play such a significant role. Weather as a "dramatic device" and "brooding force", as Peter Moore has so rightly called it,[6] is artistically depicted in Shakespeare's plays; the shipwreck in "The Tempest", the fogs in "Macbeth", King Lear in the storm, and many more such scenes act as forces with their own intrinsic value and power in the drama. Using a rich variety of metaphors, weather in Shakespeare's plays serves, in a kind of early modern artistic continuation of the older religious code, to depict the human state. Obviously, the very cold winters that Shakespeare experienced in the Little Ice Age have also left their traces in the weather metaphors of his plays.[7] The poet explicitly connects ageing and the seasons, as well as love and sunshine: "age is as a lusty winter, Frosty, but kindly",[8] and "Love comforteth like sunshine after rain".[9] Obviously Shakespeare's dramatic world develops on the threshold between older religious weather representations and beliefs and modern meteorological modes of weather observation. Nevertheless, he does not seem to leave the one for the other but rather experiments with modes of reconciling subjective and rational, religious and modern, and moral and meteorological weather.[10]

The famous Swedish theologian Gustaf Aulén may also have drawn on Shakespeare's art when he, in his influential historical-systematic interpretation of the long-term history of salvation, circumscribed this as a drama of salvation.[11] What role did (and will) weather play as a device in the drama of historical and ongoing salvation of Creation?

Extreme weather as a response to social injustice and lack of love for the poor

Over a thousand years before Shakespeare, the Cappadocian theologian Gregory of Nazianzus (329–390) reflected on weather in the context of extreme weather. For him and his contemporaries, this was extremely challenging, leading to misharvests and famines, causing suffering, especially among the poor. Their exploitation by the rich was heavily criticised by the theologian, especially because landowners went so far as to increase the scarcity of food by holding back the limited amounts of seed in order to achieve higher prices and profits.[12] For Gregory, the weather that damaged the crops and caused the insufficient harvest was an act of God intended to morally shake the rich with regard to their unjust practices towards the poor. In Gregory's view, God was preaching through creation, and weather was for the Creator a pedagogical medium for educating believers.

In his 16th homily, Gregory departs from the experience of a bad hailstorm that damaged the crops of the inhabitants of Nazianzus in 373 AD. While his father had kept silent about it, Gregory feels the need to formulate a quick answer in his sermon about the deeper causes of this plague. After a cattle plague had devastated the herds, a long drought threatened the harvest, and finally the crops were ruined by a storm of rain and hail. In a long elaboration of many biblical sources, mainly from the Hebrew Bible, Gregory analyses analogies for such extreme weather which all are located in a relation between God the Creator and his people.

The painful question, both for the local farmers and for the theologian, is,

> Whence come famines and tornadoes and hailstorms, our present warning blow? . . . And how is the creation, once ordered for the enjoyment of men, their common and equal delight, changed for punishment of the ungodly, in order that we may be chastised through that for which, when honoured with it, we did not give thanks, and recognise in our sufferings that power which we did not recognise in our benefits?
>
> (Or. 16.5)

The answer to the "whence" lies, according to Gregory, not in "some unreason of the universe" but in the state of the social community of the humans. No doubt for him this extreme weather is God acting to punish and chastise. God's power over life is revealed not only in benefits but also in challenges and disasters. Gregory quotes Psalm 79:4: "We have sinned, and Thou art angry". Nevertheless, anger is, according to the theologian, not natural for God, but rather mercy. He is "forced" to anger by humans (Or. 16.14). Towards the end of the speech, Gregory no longer speaks in general terms, with biblical references about sinners who have provoked the Creator, but stops mincing his words: it is the practice of social injustice that has led to the crisis. In a way that is characteristic for the Cappadocian theologians of that time, he locates the ethical role of the church in the challenging context of economic crisis, in line with the key principle of love for the poor. Gregory even coined a specific Greek term, φιλοπτωχία (philoptōchia, love for the poor), to underline that the command to love God and neighbour was best fulfilled and practised by loving the poor.[13] The experience of extreme weather therefore demanded a greater love for the poor.

For the inhabitants of Nazianzus struggling with the damage caused by the hailstorm, Gregory identified the reasons for the extreme weather in detail, locating it in social injustice (Or. 16.18): the poor had been oppressed and wrested from God's land; neighbours had been robbed; "another man defiles the earth with usury and interest, gathering where he did not sow and harvesting where he did not plant".[14] "God was robbed" of the first fruits, others had "no pity on the widow and orphan", and others had "turned aside the way of the meek, and turned aside the just among the unjust".[15]

The theologian moreover accuses of self-righteousness those who say "Blessed be the Lord, for we are rich". According to Gregory, the disastrous weather has its cause directly in this lack of concern for justice and the poor:

> Because of these things the heaven is shut, or opened for our punishment; and much more, if we do not repent, even when smitten, and draw near to him, who approaches us through the powers of nature.
>
> (Or. 16:18)

Only atonement can be the answer in this situation: to "break their bread with the hungry, to gather together the poor that have no shelter, to cover their nakedness and not neglect those of the same blood" (Or. 16:20). Atonement will stay the plague, and "so doing, thou wilt make God to be our God, wilt conciliate heaven, wilt restore the former and latter rain" (Or. 16:20).

We do not, of course, know if Gregory's advice led to the mitigation of the extreme weather in Cappadocia. One can expect and imagine that the believers who had repentantly gathered in the church that his father had built were following his "thunder preaching" and contemplating their own practice of injustice. Even if the classical scholars had acquired a high degree of knowledge about nature and the universe, and also dealt to some degree with meteorology, the process of weather change was never regarded in a purely rational and scientific lens. On the contrary, extreme weather was, as we know from many sources, always experienced and interpreted as a consequence of human social activity and a course that had to be adjusted. Bad weather offered a sign from the powerful Creator that something was out of order in the human community.

Even if in modern times not many among the scientifically literate would ascribe to extreme weather such a direct religious connection to human activity, the lesson learnt at present about climate change can serve as a reminder. Is bad weather, now on local as well as on global scales, really anthropogenic, and whence does it come if not from an unsustainable mode of production and consumption in the global human society? Does the request for repentance and atonement still make sense, and can it turn into a stronger emphasis on love for the poor, spelled out as environmental and climate justice?

In Gregory's time, one could become aware of the interconnections between the economic, ecological, and political decline of his day, as I showed earlier.[16] The mode of how he projected all this onto the screen of weather, where he recognised in the events of a dynamic sky the powerful, angry, and merciful God, might feel strange and primitive today. But is such an elementary cultural code of regarding nature as a mirror of society really so strange? Or might it, in Goethe's words, still make sense that it is "only in the world" that we become aware of ourselves as human?[17]

Weather as an arena for the demons or a canvas for God's good governance?

Nevertheless, not only could demanding and threatening weather in former times be caused by human sin, but it was also postulated to have its roots in the activity of demons. Scholars of the 19th century postulated that Thomas Aquinas had asserted this as a dogma of faith, despite the fact that he never formulated such a dogma in the *Summa* or in any other text.

Rather, the *idea* that the medievals believed in demons' influence on the weather seems to have been paid particular attention in the 19th century, as expressed, for example, by Andrew Dickson White:

> Rains and winds, and whatsoever occurs by local impulse alone, can be caused by demons. . . . It is a dogma of faith that the demons can produce wind, storms, and rain of fire from heaven.[18]

Within this fabricated and untrue image of Aquinas' dogma of faith, it was furthermore believed that consecrated bells were able to resist the "atmospheric mischiefs of the devil". According to the Swedish poet Viktor Rydberg (in his *Medeltidens magi* [The Magic of the Middle Ages] from 1896), they had the capacity to repel demons and to avert storms and lightning.

Probably the historians of the 19th century have confused the medieval (and later) practice of ringing bells to avoid damage from storms with theological arguments about creation in Thomas' *Summa*.[19] Rather than ascribing threatening weather to the power of demons, we can expect Thomas to continue the tradition of ascribing power over changing weather to the Creator directly. Otherwise, we would have theologically established a conflict between weather-making demons on the one hand and the Almighty who could both comfort and punish through weather on the other. The biblical sources for belief in such a weather-governing Creator, which Thomas and other medieval theologians were very well aware of, were far too many to be replaced by belief in demons. A short, by no means comprehensive, list might equip the reader with some feeling of the strength of divine weather governance in Jewish-Christian faith:

- It is God, not Satan, who controls the weather (Exodus 9:29; Psalm 135:6–7; Jeremiah 10:13),
- God controls the skies and the rain (Psalm 77:16–19),
- God controls the wind (Mark 4:35–41; Jeremiah 51:16),
- God upholds and sustains the universe (Hebrews 1:3),
- God has power over the clouds (Job 37:11–12, 16),
- God has power over lightning (Psalm 18:14), and
- God has power over all nature (Job 26).

Undoubtedly, weather in classical and medieval times represented a dramatic space where the struggle between good and evil took place. In the earliest days of the Church, Christians could be accused by their opponents of having caused bad weather, and of having offended the gods, who reacted by sending bad weather.[20] Weather functioned further as a kind of canvas where God painted his emotional self and revealed it to created beings. The Jewish-Christian code of perceiving weather as a kind of divine canvas and moral barometer dominated European history until the 18th century. Hoffmann talks wonderfully aptly about the meteorological phenomena as "translators of the divine mood towards the humans".[21] Only later was a systematic exploration of weather phenomena initiated. Meteorology had certainly already been established in antiquity, but in Aristotle's famous work, and its reception by, for example, Theophrastus and later Thomas Aquinas, meteorology was, as we will see in the next chapter, still mostly an intellectual and speculative discipline where one could investigate and learn about the general process of nature as a God-given or divine and astronomical revelation of the one single spiritual system that lay behind the complex system on Earth.

The theologians of the Early Church followed the Greek philosophers in their views and seem to have drawn only marginally on biblical sources. Aristotle in particular had a great influence on them, stronger in the East than in the West.[22] Weather complexity and change was not a focus of interest at that time, but rather weather revealed the unity of the reigning spiritual powers within and beyond it. In the 18th century, for the first time, meteorology developed into an independent system of exploration, which we will mine more deeply later. Until then the religious code was dominant, in which weather in general remained a mystery and where it was perceived and interpreted as a sovereign expression of God's salvation history. The flux of wind and weather thus offered a canvas for a painting Creator, a stage where the drama of good and evil was directed, a brooding force, and a moral barometer that showed the moral state of society. Weather appeared as spiritual and moral weather, as a screen where the Creator physically manifested his/her interaction with creation and its created beings. It served, further, as a pedagogical instrument to educate those who were created in his image to stay on or to return to the right path of social and environmental justice, in the frame of love towards God, oneself, one's neighbour, and the poor.

Two exciting examples of how Christian faith communities have used weather for the interpretation and explanation of natural and sociopolitical challenges are found in Christian Europe in late medieval and early Protestant times. In the following, we will take a closer look at the so-called *weather witches* and German Protestant *thunder preaching*. Here the example of the accusation of weather witchery represents a serious and fatal lapse in the functioning of the dominant code, namely that it is God who is in control of the weather and not humans or demons. The example from two relatively unexplored documents (*Bronteiologiké* and *Teratologia*) shows, by contrast, how the traditional code of the weather as God's canvas was practised in a regional context of demanding and threatening environmental change in the 17th-century German countryside.

Weather witchery

Hans Baldung's famous woodcut *Witches* from 1508 (Fig. 4.1) strikingly illustrates the belief in the power of weather witchery that was common in late medieval times.

A group of obviously mature and experienced women are gathering in the landscape. They have taken off their clothes – as naked practising was more appropriate for impacting on weather[23] – and are busy with some kind of ritualised practice where specific artefacts are in use. Decorated bowls, long wooden bars, textiles, brushes, plates with animals, and more seem to have some kind of magic use, though the method and purpose remains hidden to the viewer. One of the women is riding on a goat through the air and seems to be doing something to the air with a hot bowl. A flow of some liquid and vapour is being produced by the two women in the central foreground. Obviously, they possess power to impact on the quality of the air. Everyone who saw such a picture at that time would recognise the magical arts of weather witchery in this image. Through its detailed insight into what the women are usually hiding from the uninitiated, one can find the explanation for the *Unwetter* (un-weather) of the time, when extreme weather painfully affected the human ecology and subsistence of the local populations across Europe and their ability to survive.

In seeking explanations of problems in the weather-dependent agriculture, where disasters, which one could not explain rationally, were striking vulnerable populations, women – especially in Germany, Switzerland, and northeastern France – were accused as witches and held responsible for the production of *Unwetter*, severe weather and disasters. The so-called weather witches show how deeply weather was understood and treated religiously at that time, but they also reveal how sexism and gender injustice caused a break with the older Jewish-Christian tradition in which one first investigated what the Creator intended to preach through demanding weather, rather than seeking immediately for scapegoats.

The practice of weather witchcraft was well known in Europe in the early Middle Ages, as one can see from documents such as Bishop Agobard of Lyon's

Figure 4.1 Hans Baldung, *Witches*, 1508, woodcut.

https://commons.wikimedia.org/wiki/File:Baldung_Hexen_1508_kol.JPG, accessed 22 October 2019

treatise against weather magic, *De Grandine et Tonitruis* (On Hail and Thunder), from 815,[24] even if the practices seem to have decreased in the High Middle Ages.[25]

Artists played a central role here in a culture where most inhabitants were illiterate. They knew in detail how this dangerous weather magic worked and they could depict and reveal it for the people. This woodcut (Fig. 4.2) shows the secret practice of affecting the weather through an offering; it depicts how animals are placed in a fire, and how the choice of species and the sequence in which they are burned was probably significant for the ritual. According to Molitor's treatise, this kind of cooking ritual aimed to summon demons through the brew.

The arts of cooking and the use of relics had a specific significance in medieval weather witchery, as shown in this woodcut about monsters and witches (Fig. 4.3).

From the iconography of weather magic and pictures like this, one can learn how weather magic took place as a social event where groups of women gathered. Juridically, it was therefore often categorised as a group offence.[26]

In analogy and contrast to biblical stories about God's ability to rule the storms, the witches were also able to impact on storms that threatened seafarers, animals, and agriculture, as is shown in this dramatic picture (Fig. 4.4).

Figure 4.2 Woodcut 6: *"Hexenküche, Hexensud" – Zwei Hexen kochen einen Sud zur Erzeugung von Hagelunwetter* ("Witches' kitchen, witches' brew" – Two witches are cooking a brew to produce severe hail), ca. 1489, in Ulrich Molitor, *De laniis et phitonicis mulieribus*, Reutlingen 1489.

Figure 4.3 Von den Unholden oder von den Hexen (About the monsters or about the witches), woodcut, in Ulrich Molitoris, Von den Unholden oder Hexen: Tractatus von den bösen weiben die man nen[n]et die hexen etc., Augspurg: Otmar Verlag 1508.

Figure 4.4 Wetterhexe, in Olaus Magnus, *Historia de gentibus septentrionalibus*, Rome 1555.

From data in the studies of the history of climate, we can clearly see a significant connection between the weather changes in the Little Ice Age and the increase in accusations of weather witchery as well as in Jewish pogroms.[27] Thus, climate change accelerated the persecutions of the witches in Europe.[28] When older women were investigated on suspicion of witchcraft, they were regularly questioned about their practices of weather-making, and their confessions, usually extracted under torture, showed a clear connection which was believed to be authentic.[29] The juridical documents from that time clearly reveal an increase in women who were sentenced to death as witches, often as weather witches. A direct connection was often suspected between extreme weather and the magic of these women, such as in a hailstorm in 1445 in Southern Germany:

> 1445. In diesem Jahr war ein sehr großer Hagel und Wind als vor nie gewesen, thät großen Schaden, ihro wegen fing man allhier etliche Weiber, welche den Hagel und Wind gemacht haben sollen, die man auch mit Urthel und Recht verbrennt.[30]
>
> (1445. This year a very large hailstorm and high winds, such as never before took place, did a great deal of damage; because of this several women were caught here who were supposed to have caused the hail and wind, and who were burned with a verdict and law.)

Hail was perceived even among the Greeks and Romans as a special phenomenon where magical forces were believed to be at work. As early as 451 BC, the Roman Twelve Tables laws punished magical theft of harvest and damaging weather magic.[31]

A more nuanced view that relativises the power of the witches could also be developed, as shown in Johannes Brenz' hail sermon from 1539. The preacher here emphasised the power of God as the one who gives permission to witches to bring about the plagues. The devil only mediates between God and humans. The hail is here explained by natural causes which nevertheless are administrated by God in order to punish the ungodly and to afflict the pious. The hail represents God's encouragement to do penance.[32]

Weather magic did not only aim at damaging others, in the form of black magic (*Schadenzauber*), it could also be performed in order to achieve fair weather. Many of these magic practices seem over the course of history to have been connected to rites of fertility, as we saw in the previous chapter. European medieval weather witches might have been regarded as inheritors of these older rites.[33] Obviously, the medieval age was characterised by a rich and locally and regionally differentiated number of practices.[34] A beautiful and eminently meaningful expression of this cultural abundance is visualised in this artistic woodcut (Fig. 4.5) where two different states of weather in a village are depicted.

To the left, one can see a couple preparing a meal in front of their house in a sunny but very dry, and probably threatening, heat that affects their crops badly. To the right one sees another family afflicted by heavy rain and a hailstorm. God the Creator dwells in a cloud and directs his blessing to the family to the right. The practices that are depicted here continue a long and rich pattern of magical action such as the use of brushes, the crossing of legs, the showing of the finger, the preparation of an offering meal, and much more. The moral message of the drawing is, however, that the family to the right practises their arts with deep faith

Figure 4.5 Augsburger Holzschnitt von, 1532, reproduced in Wilhelm Gaerte, *Wetterzauber im späten Mittelalter*, Rheinisches Jahrbuch für Volkskunde 3. Jahrgang, Bonn 1952, 227.

and is blessed by God, while the family to the left relies on the magical forces, and will lose control.[35]

We may wonder if the overarching code of the Creator holding power over weather was radically set aside in the witch burnings, or if the women were rather forced to act as scapegoats in order to keep the same idea functioning, that is that extreme weather always represents a reaction of God to sinful human beings. And as they could not locate within their own community the failures that could explain the plagues, the community was forced to look for specific individuals to blame, who were found in suspicious older women and sometimes also among the Jews.

As early as the High Middle Ages, it was questioned if weather magic existed at all.[36] Argobad of Lyon and Burchard of Worms regarded it as nonsense, and Agobard argued against the belief in weather magicians who lived in a sky world called *Magonia* and used flying ships to throw hail on the ground below. Furthermore, he criticised the *tempestarii*, the persons who pretended to control weather and storms by communicating with the aerial sailors on the flying ships and whom one paid when collecting crops.[37]

The reasons for the persecution of women as witches are, of course, complex, and a profound historical investigation has considered these from many angles. Nevertheless, the strong relation between environmental challenges in weather change, the violent acknowledging of women for these changes, and their punishment in the belief that this could change weather for the better offers a clear picture.

At the same time, other violent social processes also took place in Europe, and it is not easy to analyse in what sense the persecutions of women, the warfare, and also the disturbances within the Reformation really interact with climate change, and especially the demanding cold periods from the middle of the 14th century.[38] One should definitely be careful with identifying one single cause of these radical social and cultural transformations. Not only negative developments can be traced in this period; the Enlightenment, the flourishing of Baroque arts, and the acceleration of modern science also took place at the same time. To what degree weather witchery might have had some part in these negative and positive developments still remains obscure.

Strikingly, Wolfgang Behringer describes the later period of the Little Ice Age as a "climate of religious pressure and violence",[39] and interprets witchery as the main crime of this historical period. Lonely older women in particular fitted in the image of a woman who had magical skills at her disposal, and the witchery could of course not only serve as the root cause of overarching extreme weather but even serve as a conflict zone where one could engage in local conflicts between neighbours. An accusation of witchery could thus easily help to eliminate an unwanted neighbour.

Not only women from lower social classes but also upper class women could be accused of weather witchcraft, especially for economic reasons. Here women were accused of practising black magic that could serve as a method for increasing the price of corn by damaging the fields of other business rivals.[40] Rich land owners often cultivated fields in different places, and unusual local weather variation, such as localised hail and rain that was interpreted as black weather magic, could be economically important when bad weather damaged a field in one place but not the other. Rich women could be ascribed the magical skill to protect land from black magic. Only those who had used their magical skills to harm and destroy were punished by law while those who used magic for the protection of their property were left in peace.[41] Magic practices that aimed at keeping away bad weather or enhancing fair weather are widely verified. One story, for example, narrates how such a practice failed when the farmers in Scheroutz and Werboutz in Eastern Europe forced all the women of the community to take a bath in order to achieve rain. Instead of the intended weather, though, one woman drowned and another became ill.[42] Whether the weather really improved that day or not is not told.

The churches also took up some of the magical practices in use at that time. In particular, the widespread ringing of bells to avoid bad weather or to impact on ongoing extreme weather, mainly storms, can be regarded as a magical practice. The consecrated bells were believed to carry significant spiritual power to impact on the flow of weather. Similar modes of belief still seem to be alive, even without a magical dimension, when, for example, all church bells rang on 27 September 2019 in the city of Malmö where the students of *Fridays for Future* demonstrated for a more serious response to ongoing climate change, or when the Green Faith Alliance rang bells every 11th day at the 11th hour for 11 minutes through 2019 in downtown Boulder, UK. Moreover, without any religious or scientific symbolism from weather witches, bells, and climatologists, people can gather against bad weather, as a Dutch protest demonstration showed in Amsterdam in the unusually wet and rainy summer of 2012, which turned into a sunny period after the event.[43]

The range of religious rituals and artefacts for impacting on the weather is broad. Weather blessings, weather liturgies, benedictions regarding strong rain and hail, processions, specifically consecrated candles, consecrated water, palm leaves, dedicated medals, consecrated wooden bars, and weather crosses belong to the instruments that have been taken into use.[44] Specific events are also remembered and located in a historical collective memory, such as in special hail celebration days where one prayed to be spared in the future.[45] Symbols such as a cross with a double crossbar, which was invented by the Jesuits, were ascribed the power to protect from thunder, storm, hail, and heavy rain. The shape stems from the Moorish region where according to the Legend of the Holy Cross two angels are believed to have delivered the cross to Caravaca in the Murcia region of Spain.[46] The shape of the *Caravaca Cross* is nowadays found in the coat of arms of the city of Caravaca de la Cruz as well as in a lot of fashion jewellery.

Blessings of weather and prayers for fair weather are still included in liturgies today, even if they obviously have lost their magical function. Margrethe Ruff interprets the ritual and spiritual resistance to extreme weather in the tradition of defending oneself against the demons, a practice that has been in use since antiquity. Even the production of noise and uncomfortable sounds belonged to such a demon defence method, in which one also could throw sticky artefacts into the sky to hurt the demons. As late as the 19th century, one could blame the authorities who had prohibited the ringing of bells for not having taken care of the environment, such that they were held responsible for weather-related damage to agriculture.[47]

In spite of rational developments such as humanism and the Reformation, the high period of witchery accusations took place, curiously, in early modern times. About 50,000–60,000 persons, mostly women, were sentenced to death and usually burned. Historians also talk about a witchery border and witch-free zones in the Ottoman Empire. In Western Catholic and Protestant regions, persecution of witches was clearly intense, while in the Eastern Orthodox and Islamic regions it was much less so.[48] Weather witchery followed the earlier discussed type of *maleficium*, that is, the evil work of a person who wants to harm another person. From 1430 onwards, one can also observe the formal inclusion of weather witchery in the category of damage witchery (*Schadenszauber*). From 1520, one can again observe a decrease in the trials against witches, which one might explain in terms of the distribution and success of the Reformation, where the doctrine of weather as a sphere of God's power and extreme weather as a sign for society to change course was preached.[49] One was, however, also able to develop an argument that could combine the doctrine of God's power over weather with the concepts of damaging witchery and therefore sentence women to death in any case, a position that was supported in both Catholic and Protestant circles. In that mode of thinking, the Swiss Reformer Heinrich Bullinger, for example, could argue that a weather witch had a pact with the devil and that in accordance with Exodus 22 she had to be sentenced to death.[50] After a calmer period, the most intense persecutions took place between 1560 and 1630. Often mass persecutions were carried out, and in the case of a hailstorm in 1562, a total of 63 witches were killed. In this period, Europe experienced an unexpected time of inflation, rough climatic change, and recurring crises of starvation and economic crises of trade and production. Thereafter, in the final phase between 1630 and 1770, the number of persecutions diminished quickly.[51]

Mauelshagen rightly criticises Behringer's simplistic method, which correlates the average temperature with the total number of executions, a method that does not offer very convincing evidence for the correlation. More detailed and locally based studies can nevertheless verify such a correlation as well as the interconnection of specific local extreme weather with the accusation of weather witchery in the juridical protocols.[52] In Norway, for example, one can find an accusation of having caused shipwrecks, storms, floods, and damage to crops.[53] An interesting correlation can be observed if one considers that those regions where many women were persecuted were also regions where weather sensitive crops, such as grapes, were cultivated. Persecutions seem to have developed out of an acute situation of crisis and misery.[54] We can probably also observe a regional difference, as weather magic mostly seems to have been practised in Southern Germany and the Alps while Northern Germany reveals far fewer cases.[55] Ecological and meteorological vulnerability seem to have been root causes of searching for scapegoats to hold responsible for damaging extreme weather. In general, one cannot, in my view, agree unreservedly with Mauelshagen,[56] who states that the relations of witchery and weather are culturally conditioned rather than climatically. Rather, one should take both dimensions into account. How did people understand the deviation from normal weather at that time? And why did they hold the women responsible? Why did they not physically adapt to changing weather instead of persecuting scapegoats? In any case we should also avoid a deterministic explanation of the accusations of weather witchery as simply due to climatic change. Rather, it seems obvious that the cultural and theological conditions and stereotypical beliefs were responsible for the emergence of accusations of weather witchery in Europe, and that these were accelerated, though not simply caused, by climatic change in the Little Ice Age.

While images of and stories about weather witches today only unfold their imaginative power in children's literature (Fig. 4.6), on the roof as a weather vane,

Figure 4.6 Book cover: Otfried Preußler, *Die Kleine Hexe*, Illustrationen von Winnie Gebhardt, koloriert von Mathias Weber © 1957, 2013, Thienemann-Esslinger Verlag GmbH.

or as a mask in Carnival celebrations, it was not long ago that they played a central role in assisting culture in adapting to unexpected environmental change.

The image of wise women gathered around a magic cauldron to cook a brew of powerful ingredients that impacted on the air and manifested a disastrous capacity to change local weather offered a fantasy of rich seductive details that could help to explain and cope with what people at that time experienced as an overwhelming force to which they were helplessly exposed.

In Shakespeare's plays, we can trace further the power of this fantasy (Fig. 4.7).

In *Macbeth*, Shakespeare artistically offers us a detailed formula used by the witches. In his world, winds could in this way be "untied . . . to fight against the churches". Trees could be blown down and castles toppled by these forces.

Let us pause and follow in detail the arts of the witches and Macbeth's response.

SECOND WITCH

Fillet of a fenny snake,
In the cauldron boil and bake.

Figure 4.7 Macbeth, IV, 3, the Witches' Cauldron (graphic)/F. Gilbert, 1859, watercolours.

https://luna.folger.edu/luna/servlet/detail/FOLGERCM1~6~6~355623~129936:Macbeth,-IV,-3,-the-witches-cauldro?sort=call_number%2Cmpsortorder1%2Ccd_title%2Cimprint&qvq=q:gilbert%20frederick;sort:call_number%2Cmpsortorder1%2Ccd_title%2Cimprint;lc:FOLGERCM1~6~6&mi=2&trs=148#, 28 October 2019, licensed under a Creative Commons Attribution-ShareAlike 4.0 International License (CC BY-SA 4.0)

Eye of newt and toe of frog,
Wool of bat and tongue of dog,
Adder's fork and blind-worm's sting,
Lizard's leg and owlet's wing,
For a charm of powerful trouble,
Like a hell-broth boil and bubble.

ALL

Double, double toil and trouble,
Fire burn and cauldron bubble.

THIRD WITCH

Scale of dragon, tooth of wolf,
Witches' mummy, maw and gulf
Of the ravined salt-sea shark,
Root of hemlock digged i' th' dark,
Liver of blaspheming Jew,
Gall of goat and slips of yew
Slivered in the moon's eclipse,
Nose of Turk and Tartar's lips,
Finger of birth-strangled babe
Ditch-delivered by a drab,
Make the gruel thick and slab.
Add thereto a tiger's chaudron,
For the ingredients of our cauldron.
. . .

MACBETH

I conjure you by that which you profess –
Howe'er you come to know it – answer me.
Though you untie the winds and let them fight
Against the churches, though the yeasty waves
Confound and swallow navigation up,
Though bladed corn be lodged and trees blown down,
Though castles topple on their warders' heads,
Though palaces and pyramids do slope
Their heads to their foundations, though the treasure
Of nature's germens tumble all together,
Even till destruction sicken, answer me
To what I ask you.

   (Shakespeare, *Macbeth*, 1611, Act 4, scene 1)

Neither Macbeth nor the people of his time received the clear answer that he demanded in the name of the dark, weather-ruling powers. Nevertheless, one can wonder if the image of the weather witches' powers almost offered his contemporaries some kind of explanation. "Why is the weather changing and damaging our economy?" humans have asked through the ages. The cultural stereotype of black magic offered them an answer: evil persons exist and they wish and are able to harm others, and some women are especially skilled at practising black magic. The reason therefore did not need to be sought in either the loving God's anger or the essence of nature as God's good creation. And it was definitely not a reason that had to do with oneself and the human community itself. The accused and persecuted women offered an easy answer. This might have been of some help for a while but could not possibly offer an answer that could assist the cultural understanding of the extreme and enhance the population's capacity to creatively adapt to it. Instead it seems to have been the classical code of God preaching directly to the sinful community through extreme weather that could cut more deeply. By holding God as the ultimate master of weather, the search for an explanation was directly thrown back on the whole community's way of functioning. Is God angry with us? Instead of women as scapegoats, one had to explore oneself critically and search for injustices and ungodly practices in all corners of one's own house. In the following, we will trace such a path into the 18th and 19th centuries.

"Gottes grosse DonnerGlocken sollen uns zur Busse locken" (God's great thunderbells shall induce us to do penance)

Witch persecutions decreased, as we saw, at the end of the 17th century, and Europe seems to have returned to what I have called here the classical code of weather faith. Even if extreme weather continued to impact on local and regional vulnerable economies, societies in the 18th and 19th centuries did not blame the women but interpreted extreme weather as a challenge for the whole community to repent, as shown in German Protestant preaching from the 18th century; we will take a close look at this in the final section.

Thunderstorms, crop failures, and droughts were interpreted as signs from God, calling for repentance. Weather thus received a deeply ethical function where the self-critical reflection of the population was encouraged by the theological interpretation. Such a code still seems to have a central function in environmental movements where the ecological crisis and accelerating climate change is understood as an alarm clock for modern society with a clear call for conversion, culturally, socially, politically, and economically. The genre of *Wetter- und Donnerpredigten* (weather and thunder preaching) appears as a kind of forerunner of environmentalism's revivalism. Weather and climate change reveal, as for example the influential students' movement *Fridays for Future* is making clear, the sins of our past and the challenges for our future.

In the following, I will dive deeper into two Protestant sermons that were given in Saxony in South East Germany by Gottfried Reinhold in 1639 and 1637. They are entitled "Bronteiologiké" (Thunder Preaching) and "Teratologia Prophētikē

De Sanguineis Igneisque Prodigiis" (Prophetic Teratology about the Wonders of Blood and Fire). Both are available as electronic documents in the University Library of Sachsen Anhalt in Halle,[57] but historians seem not to have paid them much attention. Gottfried Reinhold (1596–1639) was a Protestant pastor of the cathedral ("Churfürstliche Sächsische Begräbnis- und Domkirchengemeinde") in Freiberg, where both sermons were given. For the theme of this book, these documents offer an exciting detailed insight into the local population's images of faith and weather and the theological interpretation of extreme weather.

While the *Teratologia* draws on God's capacity to impact on the weather by turning the water of the Nile into blood, and dividing the water in the Red Sea for Israel's escape, and then emphasises the need to believe in miracles (Wunder) and that only God can perform these (5–6), the *Bronteiologiké* refers to a specific weather event, which is a thunderstorm that took place in Freiberg at precisely the time of the noon sermon on the day of Ascension.

The *Teratologia* refers to miraculous signs (Wunderzeichen) similar to those in biblical Egypt. These have also taken place as signs of blood and fire in Germany in the present year (6). After a long description of biblical examples in the books of Moses, the Prophets, and the Psalms, Reinhold interprets the miracles of water turned into blood, fire, and pillars of smoke (Rauchdampf) as God's expression of anger. Here it is not the unpredictable weather flux that is in view but the violence of the war that calls for repentance and atonement.[58] In his own time the preacher recognises the signs of blood and fire:

> In diesen Blutigen und Fewrigen Zeiten
> da Blut vergossen wird wie Wasser
> und die Länder fast gantz außgebrennet werden. (17)

Blood and fire are relevant for us; they are God's signs and they demand atonement (18). Strongly Reinhold attacks the "bloodhounds" who spill innocent blood, are worse than pagans, "und des Kriegsbluts nicht wollen satt werden" (and who cannot be sated with the blood of war) (19).

Atonement is also at the core of Reinhold's thunder preaching. The document's front page compresses its message eloquently:

> Gottes Grosse DonnerGlocken
> Sollen uns zur Busse locken.
> (God's great thunderbells
> shall induce us to do penance.)

With an allusion to Job 5 the preacher clarifies the intention of his sermon, entitled *On Thunder and Violent Storm* (Vom Donner und Ungewitter) (4):

> Gott donnert mit seinem Donner grewlich
> und thut grosse Ding
> und wird doch nicht erkandt. (4)

(God thunders dreadfully
and does great things
and is still not recognised.)

The aim of the sermon is not just that the congregation perceive the extreme weather but that they also recognise God within it. After some biblical examples about how God has chided sinners, including through horrible plagues, Reinhold identifies God in the weather:

Es schilt aber auch Gott der Herr im grewlichen Donner und Ungewitter. (5)
(But God the Lord also chides in the dreadful thunder and violent storm.)

The frightening thunderstorm of last Ascension Day also offers a strong reprimand. It appeared over the city, and frightened the inhabitants with darkness, high winds, hail, and flashes of lightning:

dadurch Gott der Herr mit schrecklichen hefftigen Donnerschlägen auff uns zugescholten
welche auch am Marckt in das Rathhaus
und am Meiznischen Thore gewaltig eingeschlagen. (6)
(thereby God the Lord is chiding us with horrible thunderclaps
which even are striking tremendously the town hall in the market square
and the *Meiznian* gate.)

Reinhold intends to explain what the thunder is (*Was er sey*), what it means (*Was er bedeute*), and what we must remember and think about (7). According to him and the biblical references, it is God himself who thunders: "Gott donnert mit seinem Donner" (7) (God thunders with his thunder). The thunder is God's voice. Such a work represents a specific revelation of the holy God's majesty and glory. In the hail, the lightning, and the high winds, too, it is God who acts (8). God dwells in the weather that reveals his power and strength (9). According to Psalm 68, his power lies in the clouds.

Furthermore, the theologian criticises the astrologers who pretend to know and to be able to predict when thunder is developing in the clouds; they sin against God's power, who alone rules in his almightiness. A lot of biblical references are quoted (Exodus 9, Jeremiah 10:51, Amos 9, Psalm 147, Job 36, Psalm 18, Jeremiah 10, Psalm 148). According to the preacher, God's intention is to frighten the human beings through his thundering, while the wild animals are also affected, hiding in their caves. Birds remain where they are and stop singing when God's thunder sounds. They are afraid of what they see and hear. God's lightning illuminates the ends of the earth (11). Hail and floods also appeared back in biblical times (Psalm 105:2, Exodus 9, Jeremiah 10:51, Book of Wisdom 5, Psalm 18:77). In addition, earthquakes have also revealed God's power. All of these have to be read as *Zornzeichen*, signs of anger (12). God is angry in his thunder. The movement of the forces in the sky is also eschatologically interpreted; they are messengers of Judgement Day (12), and they are messengers of war (13).

Thunder represents for Reinhold God's *opus magnum*, his great work (14). Extreme weather is when big things happen. But even God's benevolence is great, as one can see in the fact that God often averts weather damage and even avoids sending fire (15).

The old tradition of ringing the church bells in times of extreme weather is interpreted by the Protestant theologian as a "floating" practice:

> Im Bapsthumb gibt man für/ihre Glocken auff den Kirchthürnen/welche sie deßwegen auch Abgöttischer teufen und weyhen/könten mir ihrem schall und klang den Wetterschaden abwenden. Derowegen lassen sie auch in grossen Ungewitter ihre Glocken leuten. Aber das heisset Gottes Donner gespottet und verschönet/drumb man es auch wol eher erfahren/daß der Donner Gottes solche Glocken und Glöckner zu boden und drümmern geschlagen hat.
>
> (20)

> (In the Papacy one believes in one's bells on the church towers/which they therefore even idolatrously baptise and sanctify/so that they with their sound might fend off the weather damage. Therefore they also let their bells ring in large tempests. But that means God's thunder is mocked and beautified/and for this we have even learnt/that God's thunder has dashed such bells and bell ringers to the ground and into pieces.)

Obviously, the old tradition of ringing the bells was broken in the times of the Reformation, where people were radically critical of all kinds of impacting on God through human behaviour. Rather than ringing church bells, one should listen to God's *Donnergeleut*, according to our source (20). "Thunder-ringing" was certainly teaching bells (*Lehrglocken*) and also bells of awakening and exhortation (21). If one listens to the word of God, one does not need to meet the God who has to speak with an angry thundering voice. The bells of thundering were therefore at the same time bells of punishment and repentance (22), and finally also bells of comfort (*Trostglocken*) (23) for the God-fearers who love the Creator and for whom all things, "including the weather", work for the best (23). God is for the theologian

> der Armen Stärke im Trübsal/Eine Zuflucht für dem Ungewitter/ein Schatten für die Hitze/wenn die Tyrannen wüten/wie ein Ungewitter wider eine Wand.
>
> (24)

> (the strength of the poor in their misery/a shelter in the tempest/shade from the heat/when the tyrants are raging/like a storm against a wall.)

The experience of encountering God in the extreme weather should, according to the preacher, lead to an attitude of gratefulness and thanksgiving. The whole city, the churches, and the schools are protected by the Creator, who keeps away the damaging weather and who protects and cares for the *Haab und Gut* (goods and possessions) (27). God is a merciful God, keeping away weather damage as well as warfare and the spilling of blood. God offers protection and comfort in life as well as in death. It is the God of glory who thunders. God is to be accorded honour and praise (28).

Reinhold does not connect the extreme weather that his town experienced on Ascension Day to anything specific in the history of the region, but he uses the thunderstorm and threatening hail to offer his believers a long and detailed elaboration of biblical weather faith. It is God himself who appears in extreme weather. It is God who uses weather as a reminder and an educational tool to address his believers. Weather is orchestrated in order to encourage humans to walk the right path. God's presence in the frightening weather makes it less frightening, as God is believed to be merciful, loving, and protecting. Even if the Creator thunders and acts in anger, those who believe have nothing to fear. In the extreme weather, they see and feel the strength and power of God. Only by faith, *sola fide*, can humans persist. Extreme weather is not just a threat; it even offers an opportunity. While thunder preaching depicts a powerful, strong, and angry God, at the same time it encourages the congregation not to be scared but to fear God in the right way, to love, to honour, and to praise God as the Creator, and also as the Lord over and within the appalling weather.

While the term *thunder sermon* represented an established theological genre at the time of the Reformation, dealing with real weather events, it is still used today, although without its meteorological dimension and only in the sense of a powerful penitential sermon or a simple scolding.

If one compares the Early Church and Gregory of Nazianzus' thunder preaching in late antiquity with Reinhold's thunder sermon in Reformation times, one must notice that Gregory's accusation of the injustice of his times appears much more sharply and clearly, while Reinhold instead searches for the spiritual causes along a generalised path of faith that he regards as not strong enough. Reinhold here seems to follow the same furrows that the Reformer Calvin had cut:

> Thunder and lightning can do nothing of themselves, but God directeth them where he pleases. If we once know this, we shall not be afraid of the thunder.[59]

In so far as the believer knows that it is God who directs the weather, he/she has nothing to fear, according to the Reformer, whereas Gregory and the Early Church theologians urged the believers to read the weather in detail and apply it as a medium of ethical God-talk to humans. One may wonder whether the ethical and self-critical capacity of interpreting weather theologically has somehow declined through the ages since the Reformation.

Gewitter im Gewissen (Thunderstorm in Conscience)

Another interesting shift takes place after the Reformation in Protestant theology and the image of the human and nature, in comparison with Catholic modes of dealing with the weather. Weather, and especially *Gewitter* (thunderstorm), moves into the inner life of the human believer. Weather shall in this sense impact on the soul and conscience of the human who is called to repent, as we could see in the aforementioned Reinhold's preaching. Weather as a threatening thunderstorm moves into the human's conscience where it does well, that is, forces the believer to repent. While many Catholic believers acted to comfort an anxious God and to limit the damage of threatening weather through sacraments, blessings, holy water, bell ringing, or amulets containing the opening words of St. John's gospel around their necks, the Protestants distanced themselves from such practices, which they considered superstition, idolatry, and *Aussenwerk* (outwork).[60] Instead they moved the weather into their inner spiritual life and preferred a faithful, pious, intense, and humble prayer, because this could "pulverise the dark and thick clouds and make the sky louder and clear".[61] Extreme weather was from this point on understood as real preaching and a call to the sinner to repent. Heinz-Dieter Kittsteiner coins the notion of *culpabilisation* to describe how weather is doing harm because of the individual believer's guilt.[62] Instead of ritual practices, the weather now has to impact directly on the believer's faith and conscience. Historians talk about the Protestant *Gewitter-Gewissen* (Thunderstorm-Conscience).[63] This moral and spiritual conscience is not only at work when the thunder rolls but needs to be continuously functioning throughout the day, including in ordinary daily life. The weather response establishes in this sense an important collective identity marker for the Protestants in contrast to the common Catholic practices of mitigation. In this process, weather turns into a central social and individual element in the life of prayer, and the communion's liturgies; faithful response to weather turns into a characteristic of Protestant identity and contributes by distancing itself from the Catholic practices. This change and distancing also explains the negation of weather witchery and rejections of beliefs about it that took place at this time.[64]

Thunder hymns

Even if the ethical dimension might have decreased, the interconnection of extreme weather and Christian faith and morality remained a vital ingredient of ordinary people's piety in the Reformation and early modern period, as one can see from the large number of hymns that deal with God's power over weather and "un-weather". Historians can identify an obvious increase in weather hymns that took place in the first half of the 17th century.[65] In the Protestant *Liederschatz* (collection of hymns) for church and home from 1837, one can find as many as 19 *Gewitterlieder* (thunder hymns) from different ages.[66]

Here, God's anger is experienced in the thundering, but thanks are also given for having been spared from damage. Nature lies in God's extended hand, and God thunders into the sinner's ear. Anger and mercy also characterise in these hymns God's work within weather. Similarly to Gregory, the hymns also interconnect human sin and God's anger in the clouds. Extreme weather offers a chance to confess one's sin and repent. Humans are scared of God's "weather voice" (3183:5); his feet are "walking upon heavy weather" and he "drives on wings of winds" (3186:2). On the other hand, even an attitude of dignity and silence can arise when the sinner's proud mind trembles:

1. Laut und majestätisch rollet
Ueber uns der Donner hin,
Bange Angst ergreift den Sünder,
Ihm erbebt der stolze Sinn.
Steht verwirrt da: Todesblässe zeichnet ihn.
2. Stille, sanfte Rue gießet,
Dieser Auftritt in die Brust,
Die den großen Schöpfer ehret,
Die sich seiner Gunst bewußt;
Kindlich Lallen,
Steigt durch das Gewölk hinauf. . . .
4. An dem schwarzen Firmamente,
Braus't das Wetter im Tumult;
Zittre, Spötter! Thue Buße!
Fühle endlich deine Schuld! Gottes Donner
Predigt dir: bekehre dich!
(1. Loudly and majestically thunder is rolling over us,
Anxious fear grips the sinner, his proud sense is trembling.
Stands in confusion: marked by deadly pallor.
2. Silence, soft peace infused this appearance into the breast,
that honours the great Creator, and that is aware of his grace;
Childlike chattering, ascending through the clouds. . . .
4. In the black firmament roars the weather in turmoil;
Tremble, mocker! Repent! Finally feel your guilt!
God's thunder preaches to you: convert yourself!)

In the same way as in classical times, in the 19th century, extreme weather is still able to reveal God in his anger, but in the despair of a meteorological disaster weather also enhances hope for the Creator's mercy. To summarise our foregoing generations' attitude simply, it makes sense to repent, to care about justice, and to pray for a natural environment where life can flourish. Faith, weather, ethics, economy, and politics are interconnected. Extreme weather events not only offer an environmental crisis for those who are affected but also reveal a continuous need for human society to care about its inner life and justice. After all, such a historical insight seems to be not far away from what the Christian churches (as we saw earlier, for example, in the Fiji islands) are proclaiming with regard to the connection between climate and justice today. The radical difference between the believers' situation in classical, medieval, and early modern times, on the one hand, and modern and contemporary times, on the other, lies in the challenge to interconnect and synthesise in depth what we know about weather, with the help of modern scientific meteorology and earth system science, with religious beliefs about God acting in and through weather and climate. The following two chapters will therefore take a constructive as well as a critical look at the science of weather, starting with its historical genesis in the classical context of beliefs and forms of knowledge.

As we will see there, the role of the weather in the conscience, the *Gewitter-Gewissen*, where God appeared as the final authority, will change so that God's authority is fading and replaced with the human and the natural. Weather does not invade the believer's faith to provoke repentance, but nature and the human being act as final authority. Conscience and weather, deeply entangled in Protestant times, are "disentangled"[67] in the Enlightenment, especially by the emergence of meteorological science. Modern Enlightenment does not need the sense of sin and the practice of repentance any longer,[68] even if one can wonder whether environmentalism is again connecting to the older cultural symbolic continuity by imagining nature as a force that humans can sin against and where environmental damage and destruction of stable climatic patterns and biodiversity are understood as human misdeeds that one must confess and renounce.[69]

Notes

1 Cf. the foregoing chapter about the "Weather God Yahweh". Kató in his study on the Book of Hosea contributes further insights into the roots of YHWH in a weather god of the type of Baal-Hadad in the Canaanite religion. Szabolcs-Ferencz Kató, *Jhwh: der Wettergott Hoseas? Der "ursprüngliche" Charakter Jhwhs ausgehend vom Hose-abuch*, Göttingen: Vandenhoeck & Ruprecht 2019.
2 Cf. Wiggins, op. cit., 151, who understands weather in general in the Hebrew Bible "as a kind of 'barometer' of Yahweh's interaction with humankind". This also fits well into Hiebert's interpretation of the Yahwist's central view where human vocation is not to manage the ecosystem "but rather to align its activity to meet the demands and observe the limits imposed by this system upon all of its members". Theodore Hiebert, "The Human Vocation: Origins and Transformations in Christian Traditions," in: Dieter T. Hessel and Rosemary Radford Ruether (eds.), *Christianity and Ecology: Seeking the Well-Being of Earth and Humans*, Cambridge, MA: Harvard University Press 2000, 135–154, 150–151. Cf. Ernst M. Conradie, *Christianity and Ecological Theology: Resources for Further Research*, Stellenbosch: SUN Press 2006, 78. Cf. also Hiebert's reflections on the significance of the climatic differences between the uninhabitable desert, the ecology of the oasis, and the regions with humid climates for dry farming: Theodore Hiebert, *The Yahwist's Landscape: Nature and Religion in Early Israel*, Oxford: Oxford University Press 1996, 61–65.
3 Severain of Gabala, *Homily* III.6, according to Immanuel Hoffmann, *Die Anschauungen der Kirchenväter über Meteorologie: Ein Beitrag zur Geschichte der Meteorologie*, Dissertation, Universität Tübingen, Tübingen 1906, 54 (later published: München: T. Ackermann, 1907).
4 Cosmas Indikopleustes, *Christian Topography* II, 129, according to Hoffmann, op. cit., 54.
5 "I have set my rainbow in the clouds, and it will be the sign of the covenant between me and the earth" (Genesis 9:13, NIV).
6 Peter Moore, *The Weather Experiment: The Pioneers Who Sought to See the Future*, London: Chatto & Windus 2015.
7 Cf. Bertram Brotherton, "Weather in William Shakespeare's Plays," *Weather* 8, 12, 1953, 361–367.
8 "While lusty winter correlates to age . . ."

> *The means of weakness and debility;*
> *Therefore my age is as a lusty winter,*
> *Frosty, but kindly* (*As You Like It*, Act 2, Scene 3).

9   ". . . love can comfort like sunshine":

> *Love comforteth like sunshine after rain,*
> *But Lust's effect is tempest after sun;*
> *Love's gentle spring doth always fresh remain,*
> *Lust's winter comes ere summer half be done*
> *(Venus and Adonis*, Stanza 132).

10 Cf. Chiari, who argues that Shakespeare reconciles the scholarly approaches of his time with popular views rooted in superstition and promotes a sensitive, pragmatic understanding of climatic events. Sophie Chiari, *Shakespeare's Representation of Weather, Climate and Environment: The Early Modern 'Fated Sky'*, Edinburgh: Edinburgh University Press 2018.

11 Cf. Gustaf Aulén's main works *Dramat och symbolerna* (Drama and the Symbols) and *Kristen gudstro i förändringens värld* (Christian Faith in God in a World of Change), where he unfolds his ideas about "The Drama of the Atonement", as the subtitle of his essay concludes. Gustaf Aulén, *Dramat och symbolerna: En bok om gudsbildens problematik*, Stockholm: Diakonistyrelsens bokförlag 1965; *Kristen gudstro i förändringens värld: En studie*, Stockholm: Diakonistyrelsens bokförlag 1967; "Chaos and Cosmos: The Drama of the Atonement," *Union Seminary Magazine* 4, 2, 1950, 156–167.

12 Cf. Sigurd Bergmann, *Creation Set Free: The Spirit as Liberator of Nature* (Sacra Doctrina: Christian Theology for a Postmodern Age 4), (with a foreword by Jürgen Moltmann), Grand Rapids, MI and Cambridge, UK: Eerdmans 2005, chapter 2.

13 Gregory of Nazianzus, *Oratio* 14: *On the Love of the Poor*.

14 Gregory (*Oratio* 16:18) is alluding here to the parable of the talents in Matthew 25:26.

15 Here Gregory (*Oratio* 16:18) is alluding to Malachi 3:8 ("Will a mere mortal rob God? Yet you rob me. But you ask, 'How are we robbing you?' In tithes and offerings". NIV).

16 Bergmann, op. cit., chapter 2.

17 Goethe, op. cit., in chapter 1, note 14.

18 Andrew Dickson White, *A History of the Warfare of Science with Theology in Christendom*, 1896, Chapter 11, 337. Cf. also Swedish poet Viktor Rydberg, who in his *The Magic of the Middle Ages*, New York: Henry Holt and Company 1879, describes how "Other demons float upon the atmosphere, causing storm and thunder, hail and snow, drouth and awful omens (whence it is said the devil is a prince who controls the weather)" (p. 12).

19 Even today one can find (on both obscure and popular scientific sites) postulates like this: "Saint Thomas Aquinas stated in his Summa Theologica, 'Rain and winds, and whatsoever occurs by local impulse alone, can be caused by demons. It is a dogma of faith that the demons can produce winds, storms, and rain of fire from heaven'" (http://miltontimmons.com/ChruchesVsLightningRod.html, accessed 10 January 2020). Nevertheless, Thomas' *Summa* does not offer any evidence for such a quote or similar belief.

20 Augustine reports in detail on this and rejects the accusation, in De civ. Dei III, 17. Cf. Hoffmann, op. cit., 31f.

21 Hoffmann, op. cit., 22.

22 For a detailed description of the patristic theologians of the West and East see Hoffmann, op. cit.

23 Erotic nakedness belonged to the common expressions of female magic in different contexts. Cf. Margrethe Ruff, *Zauberpraktiken als Lebenshilfe: Magie im Alltag vom Mittelalter bis heute*, Frankfurt am Main and New York: Campus 2003, 49f. Cf. also

Gerhard Gesemann, *Regenzauber in Deutschland*, PhD Dissertation, University of Kiel, Kiel 1913, 18f., who describes in rich detail how the sources report groups of women who practised naked in order to achieve fair or prevent bad weather.

24 Cf. Rob Meens, "Thunder Over Lyon: Agobard, the Tempestarii and Christianity," in: Carlos G. Steel, John Marenbon, and Werner Verbeke (eds.), *Paganism in the Middle Ages: Threat and Fascination*, Leuven: Leuven University Press 2012, 157–166.

25 Christian Rohr, *Extreme Naturereignisse im Ostalpenraum: Naturerfahrung im Spätmittelalter und am Beginn der Neuzeit*, Köln Weimar: Böhlau Verlag 2007, 430.

26 Heike Albrecht, *Hexenglauben, Hexenverfolgung, Hexenwahn im Deutschland der Frühen Neuzeit: Ansatz einer soziologischen Analyse* (Magisterarbeit an der Universität Kassel), Hamburg: diplom.de 2001, 51, www.diplom.de/document/221050.

27 Rüdiger Glaser, *Klimageschichte Mitteleuropas: 1200 Jahre Wetter, Klima, Katastrophen*, 3rd edition, Darmstadt: WBG 2013 (2008), 196. Cf. also Susanne Kiermayr-Bühn, *Leben mit dem Wetter: Klima, Alltag und Katastrophe in Süddeutschland seit 1600*, Darmstadt: WBG 2009, 60ff.

28 Wolfgang Behringer, *A Cultural History of Climate*, Cambridge, UK and Malden: Polity 2010, 128.

29 Glaser, op. cit., 12.

30 Bayerische Staatsbibliothek, Cgm 2008, quoted in Glaser, op. cit., 91; cf. 117.

31 Ruff, op. cit., 64.

32 Kiermayr-Bühn, op. cit., 61.

33 Cf. Ruff, op. cit., 108.

34 Cf. elaborately Gesemann, op. cit.

35 The image is analysed in detail and discussed within its historical context by Wilhelm Gaerte, *Wetterzauber im späten Mittelalter* (Rheinisches Jahrbuch für Volkskunde 3. Jahrgang), Bonn 1952.

36 Ruff, op. cit., 65f.

37 Cf. Ruff, op. cit., 109; and Meens, op. cit., 159.

38 Frank Sirocko (ed.), *Wetter, Klima, Menschheitsentwicklung: Von der Eiszeit bis ins 21. Jahrhundert*, 3rd edition, Darmstadt: WBG 2012 (2009), 189.

39 Behringer, op. cit., 116.

40 Johannes Dillinger, Thomas Fritz, and Wolfgang Mährle (eds.), *Zum Feuer verdammt: Die Hexenverfolgungen in der Grafschaft Hohenberg, der Reichsstadt Reutlingen und der Fürstpropstei Ellwangen*, Stuttgart: Steiner 1998, 79.

41 Ruff, op. cit., 110.

42 Ibid., 115.

43 www.nextnature.net/2012/07/dutch-protest-against-bad-weather/, accessed 1 October 2019.

44 Ruff, op. cit., 116.

45 Kiermayr-Bühn, op. cit., 56f.

46 Ibid., 57.

47 Ruff, op. cit., 118.

48 Mauelshagen, op. cit., 107.

49 Ibid.

50 Bullinger, referred to in Mauelshagen, op. cit., 108.

51 Mauelshagen, op. cit., 109.

52 Ibid., 110f.

53 Ibid., 111.

54 Ibid., 112.

55 Ingrid Ahrendt-Schulte, *Weise Frauen, böse Weiber: Die Geschichte der Hexen in der Frühen Neuzeit*, Freiburg: Herder 1994, 38.

56 Mauelshagen, op. cit., 113.

57 Gottfried Reinhold, *Bronteiologiké. Aus dem 5. Vers des sieben und dreissigsten Capitels des Buchs Hiobs. Als es am Fest der Himmelfahrt Christi zu Freyberg unter der Mittagspredigt eingeschlagen / Gehalten . . . Durch Gottfried Reinholden Prediger*

daselbst. Im Jahr M.DC.XXXIX, Freybergk: Beuther 1639, Halle, Saale Universitäts-
und Landesbibliothek Sachsen-Anhalt, 2008, online edition:
  http://digitale.bibliothek.uni-halle.de/vd17/content/titleinfo/469768.

Gottfried Reinhold, *Teratologia Prophētikē De Sanguineis Igneisque Prodigiis: Aus
heiliger Göttlicher Schrifft der Gemeine Gottes vorgetragen / in einer Christlichen
Predigt / am XII. Sontag nach dem Fest Trinitatis / Durch Gottfried Reinholden/ Predi-
gern in der Churf. Sächs. Begräbnüs- und DomKirchen zu Freybergk. Im Jahr Christi
1637*, Freybergk: Beuther 1637, online edition Halle, Saale: Universitäts- und Landesbibliothek Sachsen-Anhalt 2008, http://digitale.bibliothek.uni-halle.de/id/469767.

58 Reinhold is referring here to the violent experiences of the Thirty Years' War in his
    region. For details, see Hendrik Heidler (ed.), *Die Deutsche Kriegschronik: Sachsen
    mit Erzgebirge*, Norderstedt: BoD – Books on Demand 2016.

59 For the Reformer Calvin, the believer has nothing to fear as he/she knows that it is God
    who directs the weather: "Thunder and lightning can do nothing of themselves, but
    God directeth them where he pleases. If we once know this, we shall not be afraid of
    the thunder" (Calvin, Sermon on Job, xxxvi.15ff.). Cf. A. Mitchell Hunter, *The Teach-
    ings of Calvin: A Modern Interpretation*, Eugene, OR: Wipf and Stock 1999, 146.

60 Stephanie Wodianka, "Vergegenwärtigter Tod und erinnerte Zukunft," in: Joachim
    Eibach and Marcus Sandl (eds.), *Protestantische Identität und Erinnerung: Von der
    Reformation bis zur Bürgerrechtsbewegung in der DDR*, Göttingen: Vandenhoeck &
    Ruprecht 2003, 115–134, 122.

61 Johann Kiesling, *Geistliches Wetter-Büchlein*, Nürnberg 1673, 9ff., quoted in: Wodi-
    anka, op. cit., 122.

62 Heinz-Dieter Kittsteiner, *Gewissen und Geschichte: Studien zur Entstehung des mor-
    alischen Bewußtseins*, Heidelberg: Manutius Verlag 1990, 54ff.

63 Wodianka, op. cit., 123; and Kittsteiner, op. cit.

64 Cf. Nadja Schuppenhauer, *Glaube, Gewissen und Gewitter im gelehrten Diskurs des
    16. Jahrhunderts* (student research paper work at the Europa-Universität Viadrina
    Frankfurt/Oder), Grin Verlag 2008.

65 Kiermayr-Bühn, op. cit., 65.

66 *Evangelischer Liederschatz für Kirche und Haus: Eine Sammlung geistlicher Lieder
    aus allen christlichen Jahrhunderten, Zweiter Band*, ed. by M. Albert Knapp, Stuttgart
    and Tübingen: Cotta 1837, Section XXXII, No. 3172–3191, 667–676.

67 Kittsteiner, op. cit., 19.

68 For a detailed and thought-provoking discussion of the change in the concept of con-
    science in the Enlightenment, see Kittsteiner's chapter *Das Gewissen im Gewitter* (op.
    cit., 25–65), where he analyses three historical periods from the primacy of the Aristo-
    telian cosmos, through the worldview's mechanisation (including physico-theology's
    responses and new confidence in the world as creation) in the second half of the 19th
    century, to the discovery of electricity and the lightning conductor where the *Gewitter-
    Gewissen* was challenged to change, and finally due to the technical advantage of
    weather management disappeared.

69 For further reflections on the question of if and how one might understand environ-
    mentalism in analogy to Christian revivalism, see the study of my former, all too short-
    lived, PhD student Tarjei Rønnow, *Saving Nature: Religion as Environmentalism,
    Environmentalism as Religion* (Studies in Religion and the Environment 4), Berlin:
    LIT Verlag 2011.

5   In suspense

Meteorology beneath the stars

In transit from Aristotelian to modern meteorology

Attempts to turn observations about the weather and its change into a system have
been undertaken for a long time. Egyptians and Babylonians were attempting to
connect weather events with astronomical processes 5000 years ago; in China and
India also, weather science was developed millennia ago.[1] In ancient Greece, pre-
Socratic philosophers, among them especially Anaximander, worked out theories
for a systematic reflection on weather events; in like manner, Thales was able to
describe the water cycle as early as 600 BC.

The most important work, however, was written by Aristotle (384–322 BC) in
his famous *Meteorologica*,[2] which became "the bible of all meteorology for the
next 2,000 years".[3] The extensive reception of this work both in the Latin- and
Greek-speaking world and in the Arabic sphere[4] was so overwhelmingly dominant
that it impacted on all kinds of intellectual thinking about the weather until the
emergence of modern science. It was taught at many universities in the medieval
period; Albert the Great and Thomas Aquinas wrote commentaries on it,[5] and the
Jesuits used it as a central source for the departure of their own advanced development of meteorological science.[6] The traces of Aristotelian and Neo-Aristotelian
meteorology lasted into the 17th century, and especially in the Renaissance, meteorology was an extremely useful endeavour, not just in science but also "for the
needs of political control, courtly life, and religious doctrine".[7]

With the invention of thermometers and barometers, one could for the first time
undertake empirical measurements of phenomena, and from Galileo onwards,
Aristotle's system was replaced with what leads into the modern science of meteorology as a study of the atmosphere. Nevertheless, Aristotle is generally honoured as the founder of meteorology, and this chapter will take its beginning in his
systematisation, which, as we soon will see, is anchored in an attempt to separate
weather from the sphere of the stars and gods.

"Meteor" depicts in Greek that which is in suspense, and for Aristotle it was
centrally important to locate meteorology in between the surface of the earth and
the orbit of the moon. While rain, winds, thunder, and lightning, as well as earthquakes, volcanoes, and comets, belonged to this sphere, the "first body", where
the stars were circling in their eternal movement, was clearly separated from

this natural sublunary sphere, where the weather was taking place. While Anaximander earlier had focused on the phenomenon of how weather comes into being, Aristotle emphasised the location of weather change within the eternity of physical matter and nature. With the help of the doctrine of the elements he constructed a differentiated system where the circling movement in an eternal course was made evident. It was important for him to separate the sphere of the stars, which were regarded as gods, from the sublunary sphere of materiality. Faithfully to Plato, Aristotle too teaches the world's eternity in his meteorology. Surprisingly, this meteorology does not aim at a teleology, although it is located in a teleological cosmology of its time.[8]

One reason for this separation was to respond to the period's new mathematical findings, which allowed the Greeks to become aware of the great distance between the earth and the stars, but the main reason was to separate the space of the uncreated eternal "first body" from the space of earthly bodies. Meteorology thus remained limited to the narrow space of physics beneath the moon.[9] The limitation of meteorology corresponded to the religious emphasis on the space of the circling stars, the deities.[10] For the philosopher, the phenomena of weather in the narrow space were therefore totally dependent on the deities and the movements of the fixed stars, which he regarded as primary.

Aristotle's emphasis on eternity was a leading principle for his theory of the spheres. In particular, the sphere of water in the atmosphere and on Earth could be analysed clearly in his system. Characteristic for him, furthermore, is the doctrine of a warm and dry telluric evaporation which he perceives in the atmosphere as well as in the depths of the earth. Even the saltiness of the sea is explained by this, according to the philosopher. Aristotle built his theories both on empirical observations and, probably mostly, on the study of literature. His meteorology was driven by overarching principles such as that the becoming took place as a circular, eternal movement.

In the words of Gregory of Nazianzus, Aristotle postulated the dominion of "stars that orchestrate our existence" and "a universal motion that controls all things".[11] This movement is again dependent on the unmoveable mover, which is the summit of all his ontology and cosmology. The earlier pre-Socratic relation of the astronomic and sublunary spheres is reduced, and a dialectic of opposites is established, such as the distinction of warm-cold and dry-wet, which both interpenetrate the world as streams. Another intention in his system is to make evident that in all the apparently unordered dwells an order. His work aims at making these eternal laws of weather change evident. Weather for Aristotle revealed the eternity and also the ordered explicability of weather flux. To this extent, one can without doubt regard him as the predecessor of modern meteorology, which mainly aims at predicting weather based on detailed insights into its "order" and the recurrence of recognisable patterns. Weather for Aristotle appears as an expression of the "Ewigkeit des Naturlaufs", the eternal course of nature.[12] Earth is imagined and explained as one planetary space of atmosphere, which according to him breathes as evaporation.

Even if Galileo and Descartes replace Aristotle's system and prepare for our modern view, Descartes can without doubt be regarded as an inheritor of Aristotle in the sense that both regard weather as involving more than weather.[13] Descartes discerns "a world cohesive and stable enough to be perceived by a sovereign gaze of the cogito; weather thus becomes an ideal subject of his studies, as a world of change and dynamism he could scrutinize and attempt to predict".[14] The path to modern meteorology now lies open.

After Galileo had constructed the first thermoscope in 1607, meteorology turned into a new paradigm where from the 17th century onwards one was able, using this as well as other instruments such as the barometer, rain gauge, and anemometer, to collect and compare precise measurements. The Renaissance certainly moved away from Aristotle's much criticised anchoring of meteorology in metaphysics. Nevertheless, its natural philosophers could still maintain some of his principles despite rejecting his thinking about final causes.[15] Aristotle, however, had not used any final causes for explanation in his meteorology but had rather talked about imperfect mixtures (of the elements). Renaissance natural philosophy and meteorology thus faced a dilemma. Did the absent teleology in Aristotle also include a statement against the causes? Probably not. Should one explain the order and aim of weather theologically as dependent on God's will and providence? Or did the unordered weather rather limit God's involvement in the world?[16] Martin shows how Lutheran theologians took a different position at that time and regarded weather signs as divine expressions of providence and anger. Rare and violent events could thus be seen as prophetic and their purpose was to foreshadow the future.[17] Weather was here interpreted directly in the light of divine providence.

Renaissance meteorology certainly distanced itself from Aristotle's metaphysics but continued on the path he had prepared. Theologians in both Catholic and Reformed camps connected the understanding of nature and weather and its teleology with religious meaning that brought ethical meaning. The final cause of contingent weather, however, was known only by God himself.[18] Aristotle's desire to describe a cyclical system of order was in this sense also preserved and developed by the theologians. The weather was a kind of a courier by which God sent messages and expressed him/herself. Faith in God was connected to knowledge about the spiritual significance of weather events, a knowledge that could offer comfort and safety. In spite of the emerging science and natural philosophy of the Enlightenment, weather was, until the end of the 17th century, still a religious issue and deeply connected to true faith.

World weather synergies and what climatically comforts us in Humboldt's *Naturgemälde*

One could perhaps claim that the 19th century first offered some kind of a de-sacralisation of weather. Scientific and technological advances have dramatically altered our perception and understanding of weather phenomena, which were explained earlier through mythology and religion.[19] In the context of the development of empirical science, both the systematic studies of the clouds and Alexander von Humboldt's geography the earth obviously played a significant role. Within meteorology particularly, his pioneering mapping of the average temperatures around the globe was a far-reaching breakthrough (Fig. 5.1). His publication of the map in 1817 can be regarded as the first step towards modern climate science. In the context of his study of the sociology of plants, Humboldt also became interested in temperature conditions, and he plotted lines, which he called isotherms, through the places with the same mean annual temperature on the map. What he discovered and visualised in this way was the distinct "azonal" component in the distribution of temperature, and hereby he laid the foundation for comparative meteorology and a method that is still used in modern climatology.[20]

Even if Humboldt through his detailed observations and collected information created the conditions for a systematic worldwide meteorology, he not only remained sceptical of forecasting the weather but indeed rejected such a possibility in general. In his main work *Cosmos*, he talked about "die problematische Vorherbestimmung" (the problematic prediction).[21] Humboldt's method organised different detailed observations along several parameters in relation to specific places. He included average calculations and interconnected all his information. In his systematic mapping, Humboldt furthermore characterised *Wetterlagen* (weather conditions) in different parts of the world and investigated their impact on vegetation, agriculture, and what he called "das Gefühl klimatischer Behaglichkeit" (the sense of climatic comfort). Both the body and the soul were affected by the weather in his synthetic view. In an exemplary way, he analysed and described what we earlier called the interconnected impacts in the atmosphere's complexity, and granted these an essential role in his overarching understanding of nature as *Naturgemälde* (nature painting):

> Der meteorologische Theil des Naturgemäldes . . . zeigt daß alle Processe . . . welche das unermessliche Luftmeer darbietet, so innig mit einander zusammenhangen daß jeder einzelne meteorologische Proceß durch alle anderen gleichzeitigen modificiert wird.
>
> (The meteorological part of the Nature Painting . . . shows that all processes . . ., which are presented by the immeasurable sea of air, are interconnected in such an intimate way that every single meteorological process is modified by all other simultaneous [processes].)[22]

Humboldt's cosmos represented a system of order and harmony that weather nevertheless exposed to a chaotic force, an insight that deeply provoked and disturbed his contemporaries.[23] Humboldt acknowledged that weather simply offers an all too extensive "Vervielfältigung und Complication der Störungen" (multiplication and complication of disorders) in the unity in totality that he was striving to explore and explain in his *Cosmos*.

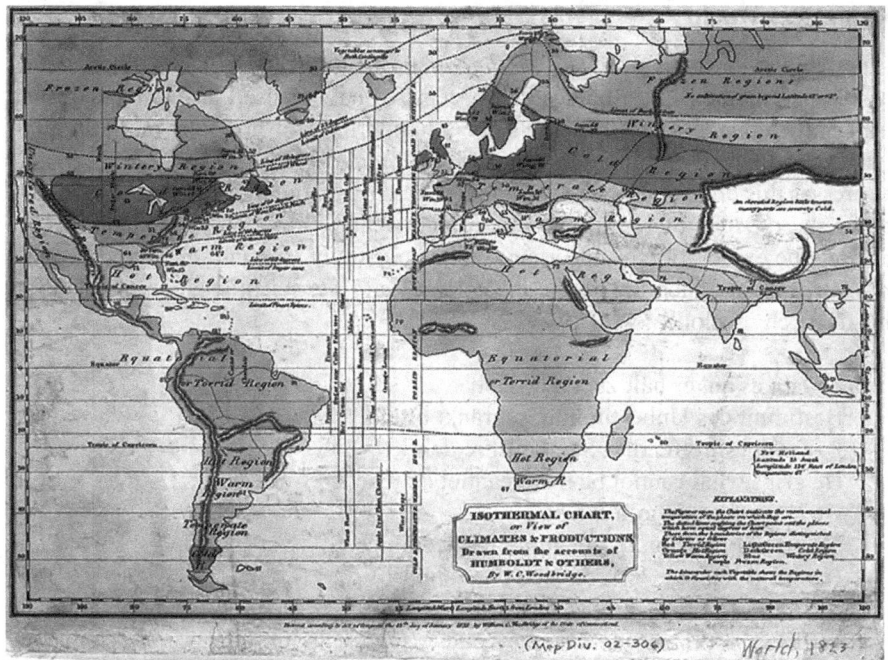

Figure 5.1 Alexander von Humboldt, *Isothermal Chart, or, View of Climates & Productions*, 1823, published by William C. Woodbridge, Hartford, CT: Oliver D. Cooke & Co., 1823, Lionel Pincus and Princess Firyal Map Division, New York Public Library. "Isothermal Chart, or, View of Climates & Productions", New York Public Library Digital Collections.

http://digitalcollections.nypl.org/items/510d47db-b00e-a3d9-e040-e00a18064a99, accessed January 25, 2018. https://upload.wikimedia.org/wikipedia/commons/7/71/Woodbridge_isothermal_chart3.jpg, accessed January 25, 2018.

In order to integrate this demanding unpredictable weather into his overarching understanding of the nature painting, he was the first to analyse global interconnections and climate patterns in his isotherms.[24] For him, weather consisted of "perpetuierliche Zusammenwirkungen" (perpetual synergies) of air, wind, ocean currents, and the vegetation cover's altitude and density on land.[25]

Humboldt even considered seriously the subjective and emotional dimension of the impact of weather on human beings. In his *Cosmos*, he defined climate as "alle Veränderungen der Atmosphäre, die unsere Organe merklich affizieren" (all changes of the atmosphere that affect our organs distinctly), and in his personal library he even had Ackermann, the standard work of his time in the field of weather and medicine.[26]

Clouds, trembling and uncertainty – science and arts in a boundless world

Luke Howard's writing on *The Modification of Clouds* in 1802–1803, where he assigns the clouds Latin names, also seems to have caused an earthquake that especially affected artists. Goethe's poem honouring Howard, who had also published the first urban climate study in his *Climate of London* 1818,[27] concisely expresses this breathtaking significance. It was still an ongoing challenge to investigate the order in the seemingly unordered weather flux, when Goethe in 1821 praised the exploration of uncertainty in his poem "*In Honour* [Ehrengedächtnis] *of Howard*", in which the heroic investigations of the scientist are applauded for his research on clouds:

> Er faßt es an, er hält zuerst es fest;
> Bestimmt das Unbestimmte, schränkt es ein,
> Benennt es treffend! – Sei die Ehre dein! –
> He grips what cannot be held, cannot be reached,
> He is the first to hold it fast,
> He gives precision to the imprecise, confines it,
> Names it tellingly! – Yours be the honour! – [28]

Today also, *uncertainty* in climate change represents an utterly new challenge for modernity. Weather change forces us to confront uncertainty. Faced with this challenge, empiricism, as an epistemology and a worldview,[29] hits a wall and collapses like a house of cards, leaving us, at best, with a new epistemological consciousness of limits and limitedness. The phenomenon of uncertainty also shuffles the cards anew in the game of power sharing. Do we need to learn to deal with general uncertainty in a different way? How can we differentiate between two kinds of uncertainty: the threatening uncertainty that must be eradicated and the uncertainty that serves as a source of insight and wisdom?

While the publications of the IPCC follow the conventional path, turning uncertainty into certainty, they come at the same time closer to the threshold at which one must radically re-estimate and re-investigate the value of the uncertain. Although divested of an older metaphysics, climate and weather are still challenging our deep attitude to unpredictability and uncertainty. Faith communities are obviously profoundly needed in dealing with uncertainty.

Howard, in a long letter to Goethe, identified himself as a Christian believer, obviously in the tradition of his father, who was a Quaker. Religion for him was not a system of notions for speculation or a series of rituals to dull one's cognitive faculty, and definitely not a system of violently forcing one's belief on others. For Howard, it revealed rather a spirit of compassion and the straight and pure path to peace and happiness.[30] Howard also held a strong belief that science would unite the nations.[31]

In handling the tremendous, the Sacred, and the invisible between heaven and Earth, religions are world champions, a wisdom that Romantic artists had preserved in times of the advancing of empirical science. As one can see in the work of Goethe, the new scientific examination of weather deeply affected his poetic work. Science and art went on walking hand in hand (Fig. 5.2). Howard's achievement, according to the poet, was

> *bestowing form on the formless, and a system of ordered change in a boundless world.*[32]

Howard's depiction of the cumulus cloud particularly enchanted Goethe when he interconnected what happens and threatens above with what trembles beneath among us.[34]

Cumulus

Und wenn darauf zu höhrer Atmosphäre
Der tüchtige Gehalt berufen wäre,

Figure 5.2 Ludwig Heß, *copper engraving*, in Goethe, *Zur Naturwissenschaft I.3*, 1807, 469.[33]

Steht Wolke hoch, zum herrlichsten geballt,
Verkündet, festgebildet, Machtgewalt
Und, was ihr fürchtet und auch wohl erlebt,
Wie's oben drohet, so es unten bebt.

Still soaring, as if some celestial call
Impell'd it to yon heaven's sublimest hall;
High as the clouds, in pomp and power arrayed,
Enshrined in strength,
in majesty displayed;
All the soul's secret thoughts it seems to move,
Beneath it trembles, while it frowns above.[35]

A key principle for the Romantic period's understanding of weather was the intertwinement of the human self and the surrounding atmosphere of weather. Goethe summarised this in his formula about knowledge of nature as indissolubly dependent on knowledge of the self and vice versa:

Der Mensch kennt nur sich selbst, insofern er die Welt kennt, die er nur in sich und sich nur in ihr gewahr wird.

The human being only knows herself as far as she knows the world, which she only becomes aware of in herself, and only in the world does she become aware of herself.[36]

Percy Bysshe Shelley (1792–1822)[37] expressed it wonderfully in the famous words of *The Cloud* from 1820:

I am the daughter of the Earth and Water,
And the nursling of the Sky;
I change, but I cannot die.

Even if modern science advanced and weather flux was step-by-step de-sacralised and freed from its metaphysical presuppositions, the Romantic era also dived into the spiritual depths of weather and emphasised it as scenery for mirroring the emotional self. In particular, the painting of the clouds received, in the aftermath of Howard's systematisation and Goethe's honouring appraisal, a central position, and the methods of depicting the sky and its clouds offered a marvellous sphere for continuing the earlier projections of belief onto the sky. Cloud paintings served as spiritual driving forces for exploring both nature and the human self. Landscape painting, and especially the detailed depiction of the atmospheres of wind and weather, attracted arts, and offered, as we saw in Turner's work, a creatively advanced new perspective on the human interactions within weathered environments.

Art historian and theologian Johannes Stückelberger, in his extensive study *Wolkenbilder* (Cloud Images), explored the enormous interest in clouds that arose from the early 19th century onwards in the arts of painting (Fig. 5.3), but also in other art forms such as film, photography, sculpture, and architecture. In our last chapter, we will also follow this concern up to our own time, when Karolina Sobecka interrelates our own perspective, standing on the ground, and the satellite's view, in orbit, on the same clouds "from both sides" (Fig. 5.4).[38]

Stückelberger differentiates between five forms of cloud images: (a) meteorological depictions, (b) images of nature, (c) metaphors (for the strange, mysterious and blurred), (d) screens for human imagination and mirrors of the soul, and finally (e) as abstractions.[39] For Stückelberger, the enormous increase of clouds in the arts mainly has to do with a new understanding of the sky, which since the beginning of the 19th century has no longer solely presented the eternal infinite but has come closer. The sky appears and is now perceived, in Friedrich Schleiermacher's words, as the infinite within the finite. Weather is not necessarily the main focus of this curiosity and intense exploration, but rain, fog, temperature, atmosphere, wind, storm, and constant and unpredictable variations appear, of course, as an integrated part of the cloud images. Insofar as cloud images also could and should be interpreted in the lens of faith and religion, one needs a subtle upset of observations and methods, which lies a bit outside of our book's theme. In general, nevertheless, one can say that arts in contrast to science have never

Figure 5.3 Caspar David Friedrich, *Abend am Ostseestrand* (Evening on the Baltic Sea), 1831, oil on canvas, 54 × 72 cm.

https://commons.wikimedia.org/wiki/File:Caspar_David_Friedrich_-_Abend_am_Ostseestrand.jpg, accessed 1 November 2019

Figure 5.4 Karolina Sobecka, *Clouds from Both Sides* (Project: The Matter of Air), 2013, by courtesy of the artist.

abandoned their intense and profound examination of the Spiritual and Sacred.[40] The artistic images of clouds in the sky (Fig. 5.5) also contribute in this way to keeping the spiritual depth of weather wisdom and wonder open, and to interconnecting it with scientific perspectives, up to the present day.

The intense interest of painters in *skying* or *clouding*, as some of them circumscribed it,[41] was coincidental with the advantages of scientific meteorology. Interestingly enough, the invention of modern weather took place both in the arts and in science at the same time. Obviously, it was deeply connected to the emergence of modernity. But only a few artists seem to have been interested in the systematic ordering of clouds as meteorological objects. Rather, most seem to have been fascinated by the clouds and the open sky as an expression of the finite within the infinite. Weather that has embraced and driven the clouds offered, in this enchanted view, an appearance both strange and familiar, both close and far away. In this way, one might wonder if both the meteorological scientist's valuation of the overwhelming complexity of the atmosphere and the artist's exploration of its finite infinity might be able to converge. My suggestion for a synthesis of wisdom, wonder, and knowledge might have been in the air since the early 19th century. Might arts, religion, and science be able to come closer to each other given the demands of changing skyscapes, weather, and climate?

After a long history of weather wisdom in the religious, cultural, and artistic spheres, modern science has gradually, in stages from the 17th century, taken over the systematic understanding of weather events. While images of the divine were mixed with observations and abstract speculations about weather flows in the Aristotelian tradition, modern science advanced without the "God hypothesis",[42] and even if giants such as Newton and Linné still regarded their scientific systems

Figure 5.5 Johan Christian Claussen Dahl, *Tordenvær i nærheten av Dresden* (Thunderstorm Near Dresden), 1830, oil on paper, 22 × 33.8 cm.

https://upload.wikimedia.org/wikipedia/commons/9/99/Johan_Christian_Claussen_Dahl_-_View_of_an_Approaching_Thunderstorm_-_Hamburger_Kunsthalle.jpg, accessed 1 November 2019

as interpretations of the work of God, along a medieval code where the book of nature reveals the Creator, meteorology developed as a sub-discipline of physics, and later also of chemistry, where the scholar de-sacralises weather as a force of God or as a direct divine revelation.

Disenchanted atmosphere

The discovery of precise technical instruments allowed, from the 17th century onwards, a steady measurement of temperature and air pressure, and later on even of winds and precipitation and humidity, and the invention of the telegraph in 1843 in particular allowed the establishment of a translocal network of weather observations, through which one could collect and compare the changes of weather on a large geographical scale. This scale was even more enlarged with balloons, computers, and more recently weather satellites that allowed extra-terrestrial monitoring of different parameters and the movements of weather patterns.

Weather has been disenchanted and has lost much of its spiritual aura even if weather forecasts on radio and TV still can nurture a nearly religious attitude in regions which are especially exposed to harsh and demanding weather. In Northern Norway, for example, one should not talk but should keep quiet in the evening when the family gathers around the weather forecast on the radio or TV. Ignoring it or not taking it seriously is interpreted as contempt. Some consciousness of the forces of unpredictable weather is thereby preserved in a way similar to that among sailors and aeroplane pilots. In ordinary culture, however, the religious attitude to weather flux seems to have shrunk to a minimum. One can only wonder if the rising consciousness of dramatic climate change as well as the emerging ecospirituality with its strong implications of neo-animism might bring a change. In green worldviews, the belief in the spirituality of non-human beings can function as an implicit force where the commitment to embody empathy for non-human life forms and to perceive the personhood of others is clothed in the language of what one could call soft animism. Might we also expect the forces of weather to be included in a holistic animistic view of in-spirited life?[43]

Our scientific knowledge about weather and climate has through the ages increased impressively, even if meteorology is scarcely taught on a more general educational scale. As a meeting point of physical and chemical subdisciplines in the larger frame of geosciences, it remains a sphere for specialists, whose applications the citizen encounters in the shape of media weather forecasts. Even if meteorological forecasts have become more precise and solid due to scientific and technical advancements, the risk of relying on them in cases when they do *not* come true increases correspondingly. Better forecasts therefore do not necessarily imply more safety.

At the centre of scientific meteorology stands the concept of the *atmosphere*. Meteorology is defined as "the study of the atmosphere and its phenomena",[44] and weather is defined as "the condition of the atmosphere at any particular time and place".[45] Weather is comprised of different elements: air temperature, air pressure, humidity (water vapour in the air), clouds, precipitation, visibility, and wind.

Climate is in this sense average weather over a long time, or a climate of a specific region. Climate represents an accumulation of daily and seasonal weather events, and it also includes weather extremes. The word "klíma" is originally Greek; it means inclination and refers to the angle of the sun's radiation on the earth's surface. It was first used by Eudoxos in the 5th century BC. In modern times, climate is defined as the totality of meteorological phenomena that characterise the state of the atmosphere at a particular place on the earth's surface; in other words, climate can simply be defined as "the average weather".[46] In a nutshell we might say that "climate is what on an average we may expect, weather is what we actually get".[47]

The atmosphere is regarded as a huge thermodynamic system that surrounds the more stable Earth.[48] Atmospheric processes take place in space and time as expansions and with structure, whereby scales serve as an important instrument for their analysis, both scales of events and scales of observation. Further modes of movement, as well as life cycles for processes that are limited in time, offer important perspectives. The theory of atmospheric circulation plays a particularly key role herein;[49] here, among other things, the uneven distribution of land and water surfaces on the planet is important. While in English "weather" is the only available term, German differentiates between *Wetter* (weather) and *Witterung* (weather condition), where the first refers to a limited time in a certain place, while the second refers to the characteristics of such weather over time.

Grosswetter is also used as a term for the characteristic weather over a large area, such as a whole region. Due to increasing appearances of extreme weather events, Germans have even absorbed technical meteorological terms such as *Extremwetter*[50] and *Starkregen*[51] into everyday language, and the German Weather Service (Deutscher Wetterdienst, DWD) has developed and established an ambitious and very well-equipped digital warning system *WarnWetter* for the country's citizens and institutions.[52]

Meteorology has developed as a specialised mathematical science, and lay people meet it in only some of its applications. What we see in weather forecasts in the media reveals merely the so-called synoptic meteorology, where a multifaceted perspective on all parameters is put together (in a *synopsis*), and a prognosis is worked out from well-known patterns of movement and change as well as from observed continuities. Long-term predictions are still totally impossible, as the weather systems are highly complex.[53] The level of change is so demanding that a long-term prognosis for more than two to three days is irresponsible even if predictions for up to ten days are given nowadays, though with reservations. The dimension of unpredictability in weather processes is, in my view, impressively high, due to the fact that the shape and development of just one cloud has the ability to change a whole series of subsequent weather processes on the border between a high- and low-pressure front.

For all the details of scientific meteorology, I can only refer the reader to the standard introductions where sections such as radiation, jet streams, global warming, earth and air impacts, and much more not only stimulate the imagination but also enrich awareness of the all-embracing complexity and continuous modes of moving between different states of weather alteration. A spiritual attitude and awareness of being alive in weather lands can without doubt also be enhanced by the rich perspectives and intertwinements of scientific meteorology, which certainly operates along the strict rational code of science without a hypothesis about the Sacred. Modern meteorology can without doubt create wonder. Nevertheless, it can open our eyes to an impressively high level of change, variation, complexity, and richness of differentiations. And it can teach us to become aware of how different impacts can be interconnected. Not only can a butterfly flapping its wings cause a typhoon far away, but chaos theory is also applicable in understanding the interconnected effects of human beings' unsustainable lifestyles on parts and the whole of the earth system.

The deep subjective experience of weather – alongside mysteries such as sleep, dreams, and romantic love – expressed by Marcel Proust's narrator, where a change of weather is "sufficient to create the world and ourselves anew",[54] can also be said to be true analogically in the scientific exploration of the same weather on the large scales of the planet and its climate complexity.

Cultural impacts, urban weather, climate history

In spite of its strict geoscientific character, meteorology is nonetheless still from time to time affected by cultural imagination, for example in the central concept of *weather fronts.* The metaphor "front" appeared in the early 1920s when two Norwegian scholars, Jacob Bjerknes and Halvor Solberg, used it as a central image to describe and analyse the movement of air masses along a course. Their *Polarfront* theory was especially widely received. With the recent experience of the First World War, in which the nations struggled in a warfare of locked fronts, in the so-called *Frontenkrieg* (war of fronts), the metaphor was uncritically transferred to the language of meteorology. Here it still serves as a current term, even if the Norwegian front theory was built on false assumptions and has since been replaced with a theory about fields.[55] Weather is, as Roland Barthes rightly points out, "not an innocent theme"; governments use it as an excuse to justify what has gone wrong, "making a mockery of the modern, technocratic State". Weather in situations of catastrophes exists, according to Barthes, only "after the event, as a discourse of irresponsibility".[56]

Another interesting exchange between meteorology and cultural and spatial studies can be found in the increasing significance that is attributed to urban meteorology. As we have already seen, Howard was curious about urban weather, and paved the way for others to develop a discipline that specifically investigates the geography and architecture of the city with regard to its impact on weather and climate, as well as its sensitivity to climate change. The city as "Wärmeinsel" (island of heat) is, especially in Central Europe, the object of intense investigation at present.[57]

In spite of the all-embracing presence of weather in human life, applied meteorology advances as a strict empirical science that does not include any consideration of the human impact, with the exception of urban meteorology. Climate impact science, which we will discuss in a later chapter, operates of course explicitly with a concept of the human within "anthropogenic climate change", even if this concept appears rather poor with regard to the human's complexity of social, cultural, and subjective dimensions. Compared to the high level of differentiation of its scientific investigations in earth system analysis, the human in this model is stereotyped and one-dimensional.

Nevertheless, an encouraging bridge between meteorology and climatology on the one hand and the environmental humanities on the other has been built by historians who have established the field of *climate history*. In its first period, climate history mainly dealt with the collecting of data about the climate's development through the ages and could include literary sources in order to reconstruct an image of earlier climate conditions. Within climatology, climate history in this sense still represents an important subfield. While climate studies in earlier times mainly aimed at the exploration of average data, climatology today has moved from the exploration of purely descriptive to causal processes, and it also tries to include human impact. Climate is now rather regarded as an earth system where variability offers a central category for analysis in the frame of thermodynamics.[58] One can illustrate the different components and impacts in a figure such as this one from the IPCC (Fig. 5.6).

Historical knowledge thus appears to be a necessary tool for observing and predicting the speed and mode of climate change. Historical climatology is developing today in a broad way and is mining more deeply the three fields of (a) reconstruction of earlier climate conditions, (b) historical climate impact research and the interactions of society and climate, and (c) the "history of knowledge" with regard to climate, that is, the history of climatology itself as well as of the perception of weather in different times. With regard to nature and society, climate history can in this way contribute to important insights into the interconnection of weather, climate, and humanity.[59]

One can only hope for a further deepening of interactions and scientific cooperation between climatology and other disciplines in the environmental humanities, and the underlying intention of this book is of course also to encourage such a new synthesis of studies of climate on the one hand and studies of religion and the environment on the other.[60] A central contribution for the humanities here could without doubt consist of a critical reflection on climate determinism, which has characterised meteorology and geoscience for a long time and still today seems to be anchored in an underlying value system in climatology. Since Aristotle, our point of departure has been the homogeneity of the climate system, with the same climatic conditions impacting on different places and times. One often obscures the active role and impact of human society and its effects on climatic change.

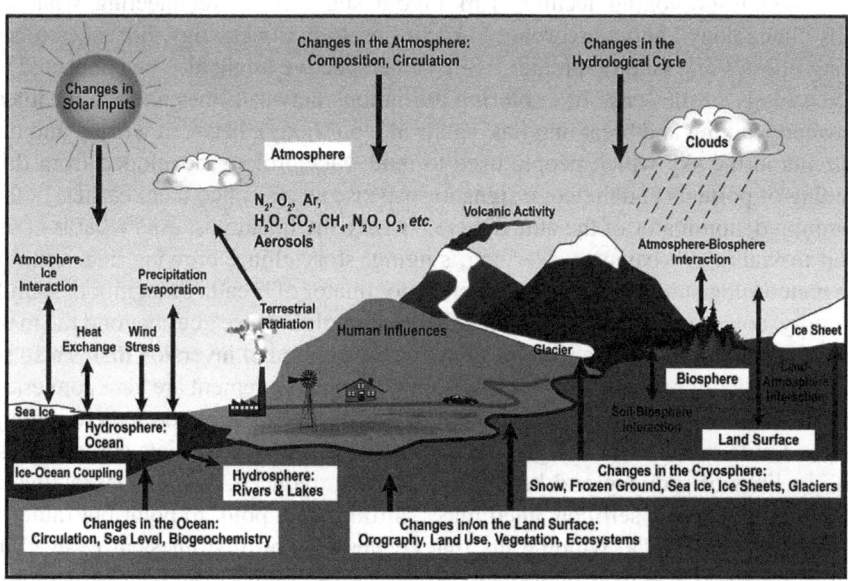

Figure 5.6 Schematic View of the Components of the Climate System, Their Processes and Interactions, FAQ 1.2: What Is the Relationship between Climate Change and Weather? Figure 1, IPCC, *Climate Change (2007): The Physical Science Basis – A Report Accepted by Working Group I of the Intergovernmental Panel on Climate Change.*

https://wg1.ipcc.ch/publications/wg1-ar4/faq/wg1_faq-1.2.html, accessed 24 January 2018

Thinkers in the humanities have also sometimes followed such a problematic climate determinism and related specific types of climate to specific types of culture, just as, for example, Japanese philosopher Watsuji Tetsuro bizarrely characterises the whole culture of East Asia as a monsoon type in contrast to the desert and meadow type.[61] Findings from climate history have already been included in our earlier discussions, and the exploration of medieval and modern European history in particular, in the lens of changing weather conditions, has produced a rich field of sources and reflections.[62]

Another critical path was trodden in this book when we departed from an understanding of weather that did not apply limited scientific-rational or an exclusively historical method, but focused on what being alive means in weather lands. Assisted by anthropology, painting, and religious studies and theology, weather has so far been depicted in this book as the temperament of being, as a mystery of atmospheric alteration, and as an expression of the Sacred, whether it be God or other spiritual forces. With regard to contemporary meteorology and climatology, I will end this chapter by discussing two further contributions that might deepen and widen our understanding of weather, faith, and wisdom.[63]

Weather as "the temperament of being"

In a thought-provoking lecture, Tim Ingold suggests interconnecting what he calls "lineaology" and meteorology. According to his lineaology, life takes place along lines of one kind or another.[64] Such a perspective might also be connected to meteorology in the sense of exploring affiliations between lines and the weather, between walking and breathing (taking air into our body), between writing and the portents in the sky which people used to read. Meteorology developed from this reading of portents and meteors, signs in suspense as we called them earlier. In the common denominator of the atmosphere, "where the linealogist asks what is common to walking, weaving, observing, singing, storytelling, drawing and writing, the meteorologist looks for the common denominator of breath, time, mood, sound, memory, colour and the sky".[65] Atmosphere in Ingold's sense goes beyond the metrics of ambient geospace, and it goes beyond the method of inversion that is also so characteristic for scientific meteorology. Growth and movement are here converted into boundaries within which life is contained. Masses are in modern meteorology mapped onto volumes, while geography compresses life into points and surfaces. Life as lines of growing and becoming is hereby reduced. A storm is, in such a view, not a coherent self-contained mass shifting from point to point but rather a movement in itself, "a 'winding up' that creates a point of stillness at its eye". It can be regarded as a whirling organism.[66] With inspiration from phenomenology, Ingold is able to depict the atmosphere as the medium of life, while weather is the term for what goes on in it. Weather – which in French derives from the word for time, *temps*, which originally in Latin denoted a mix – is for Ingold a mixture in time. The time of weather is for him a time without history, pure fluctuation.

> There *is* a pattern to the weather, but it is one that is continually woven in the
> multiple rhythmic alternations of the environment – of day and night, sun and

moon, winds and tides, vegetative growth and decay, and the comings and goings of migratory animals.[67]

Beings that inhabit the earth also inhabit the air. Without air no one could live. Therefore, the quality of our interaction, for which the medium is the condition, is tempered by what is going on in the medium, that is, the weather. Therefore, the weather is, for Ingold, the temperament of being.[68] This seems to correspond to what I earlier called the all-embracing present everywhere and everywhen. In the process of inversion, scientific meteorology has brought the weather outdoors into our constructed world indoors. Against the atmosphere of meteorology one can contrast the understanding of atmosphere immersed in phenomenology. Discussing Benjamin's "aura", Binswanger's idea of a "gestimmter Raum", a space of mood, Bollnow's unity with surroundings, and Böhme's neither-subjective-nor-objective atmosphere, Ingold rightly notes the absence of weather in these philosophical approaches.[69] This leaves us with a situation where atmosphere means one thing in meteorology and another in phenomenology, and where both disciplines assume certain truths of the atmosphere without entering a dialogue about it with each other.

Following Merleau Ponty, Ingold suggests taking literally the philosophers' "inspiration and expiration of Being".[70] Breathing the air is, then, perceiving *in* the air. All perception then depends on the air and thereby the weather. "To perceive things, then, is simultaneously to be perceived *by* them: to see is to be seen, to hear is to be heard, and so on".[71] The alternation of inhalation and exhalation marks, according to Ingold, "a *time* that is kairologically attuned to the rhythms of the environment and enacted in the weather-wising of its inhabitants".[72] Being alive, for Ingold, means living along the meshwork of lines in exhalation and being immersed in the atmosphere on inhalation. The relation of lines and the weather leads to a transformation, for example where the heaving lungs of the ploughman become furrows in the earth. For Ingold, this relation is fundamental to animate life.[73]

Weather's "technological writeability"

While Ingold does not get lost in detail in criticising empirical meteorology, environmental sociologist Bron Szerszynski radically questions scientific climatology and "expose[s] modern scientific practices of reading the climate as already containing within themselves a notion of weather's technological writability".[74] His aim is to find an alternative way of reading and writing the weather, one that takes place in the opening of climate. The contemporary story of anthropogenic climate change all too easily moves from reading to writing the weather, and it implies the solutions, from science to technology, and answers its own questions. According to Szerszynski, the dominant narrative about diagnosis and (technological) cure diverts us away from the very possibility of appropriate response.[75] Szerszynski instead pleads for grounding our understanding of climate change in an analysis of the evolution of the human metabolism with nature. He furthermore deconstructs the reading of climate technics and questions the approaches operating with a given weather technological writeability. Regarding the metabolism of human and natural, biosemiotics in von Uexküll's sense and Derrida's concepts of *différance* and *dissemination* serve as a valuable method for exploring how metabolism takes shape *within* Gaia's body. Furthermore, Szerszynski analyses climate technology in two ways: one as a rise of a green technic-economic paradigm, and another in the technocratic mandate for geoengineering intervention in the earth's metabolic system.[76] Weather hereby "has been pulled towards a certain kind of reading that constitutes it as a code that can be mastered and controlled".[77] The metabolic nature is simply understood in causal terms where climate change offers a problem that can be solved rather than an opening to be responded to. Following Jankovic's detailed analysis of the history of weather wising and meteorology in Britain, Szerszynski describes how older practices of weather wising and elite meteorology have been suspended by modern meteorology, which mirrors the change towards an industrial society, where weather wising, local knowledge, and religious hermeneutics are undermined.[78] Meteorology in a scientific mathematical and empirical key has thereby prepared the way for a technocratic weather and climate writeability that dominates the discourse on climate change at present. Nevertheless, not only is the weather errant, but "*Writing itself* (as the condition of im/possibility of meaning) is always aberrant, and reading the climate is thus always already subjected to the vagaries and aporias of writing".[79]

As a path out of this, Szerszynski suggests an opening of climate, in order to overcome the dominant technological framing of climate change that in his view ultimately constitutes a more radical evasion of responsibility.[80] How can scientific weather, which has brought weather indoors, allow us to experience climate change? Reading the weather in the open, released from its technological incarceration, would for Szerszynski also be to recognise both reading and weather as aberrant.[81] Meteorology as meteorography represents a practice of inscription in which meaning and reference can never be finally stabilised. The weather is in Szerzsynski's view a "pure medium, the region of mixing of earth and sky, the taking place of life, of metabolism".[82] Weather in this sense reminds us continuously about the openness of existence and the impossibility of autonomy. If weather is regarded as a book, it can also be regarded as writing itself. In that case, we would end with the metaphor of the book of nature that has been significant for the whole of Western history. The human, Szerszynski concludes, would then pass away as an "*archon* of nature".[83] The way to an alternative understanding of nature, weather, and also faith would then be open. "Life in the open air" summarises the human condition of responsibility towards climate in the Anthropocene.[84]

Towards a new synthesis of weather, wonder, and wisdom in the face of human vulnerability

Following our foregoing reflections about being at the mercy of weather alteration, as taught in Turner's painting and in Romantic cloud images, and deepened by our critical elaboration of the history of science, we can now, sharpened by Ingold and Szerszynski, become aware of how modern meteorology includes a

fatal danger of perceiving weather and climate simply as a measurable, mathematically readable, and technowriteable phenomenon that is under human control
and that can be mastered by economic and engineering methods. Obviously it is
not. Applied meteorology also, in working out detailed forecasts, admits that the
complexity of the weather (as a system) is so high that a very few, and presum-
ably small, events can impact radically on a whole series of subsequent events.
Without doubt one can also from within (indoor, computer-connected) theoretical
meteorology achieve a position of respect for the unforeseeable forces of weather.
Nevertheless, scientific meteorology is anchored in the codes of the industrial and
modern, including late modern, understanding of the world as a commodity, and
contributes to the fetishisation of life as a gift. Does science also perform as a
"modern mythopoesis"?[85]

Alternative ways of approaching weather as life in the open air and regarding it
as a temperament of being are possible. Earlier historical modes of weather wising
that were mixed up with elements of faith, local wisdom, and cultural skills provide, together with such opening perspectives in environmental anthropology and
sociology, a much more wide-open possibility of perceiving weather as an essential force that makes life possible, rather than an atmosphere under our control.

Atmosphere in the phenomenological key is something that takes place in
between. It happens in between nature and society and in between the subject
and his/her surroundings. Weather that permanently entices and embraces gives
life, which it then nurtures, and it challenges human awareness, remembrance,
and imagination. Faith-based attitudes and practices towards and within weather
allow encounters with the divine, not beyond but within such weather. God reveals
him/herself as weather, as the life-giving Spirit in the new shared world to come.
How one might achieve an integrated understanding of meteorological and emotional atmospheres in and beyond the age of humans, the Anthropocene, will be
discussed later in the final chapter.

Concluding preliminarily the discussions of this chapter, it should be clear why
we should not leave scientific indoor meteorology to the ruins of modernity but
rather scrutinise its ideological claim to power over the full explanation of what
weather and climate are. The challenge, in my view, is not to simply instrumen-
talise meteorology in the modern way but to reconstruct and relocate it within
a wider, religiously driven awe of and respect towards the weather and its flux.
Could one merge methods from meteorology with wisdom about being alive in
weather lands? Would wisdom then be able to be continuously anchored in wonder and awe, and at the same time offer methods for evaluating the use of knowledge? How can one open weather and climate to such a new synthesis?

Such a vision of a synthesis appears as even more urgent if we take into account
the increasing violence that is used against the poor and vulnerable populations of
the world in different forms of climate injustice. I am well aware that my reflections about the pros and cons of modern meteorology are also characterised by my
context in the Western perspective. The synthesis envisioned here is still in deep
need of impact from, and exchange and amalgamation with, other cultural spheres
of the world as well. Might it be that a promising point of departure for such a

transcultural, translocal, and transhistorical synthesis of weather wonder, wisdom, and science is found in the *vulnerability* of human life?[86] While some are much more vulnerable and exposed than others to the dangerous impact of a changing climate (mostly caused by the less-vulnerable populations), we are as humans nevertheless capable of reducing and finally eliminating the asymmetry of global climate injustice. A new transcultural synthesis of weather wonder, wisdom, and prudence can by no means in itself achieve such a global equity with regard to climate-justice-asymmetry, but it might generate one among several constructive driving forces and energies for such a great transformation.

Notes

1 Helaine Selin (ed.), *Encyclopaedia of the History of Science, Technology, and Medicine in Non-Western Cultures*, Berlin, Heidelberg and New York: Springer Science & Business Media 2008, 1662–1666.
2 Aristoteles, *Meteorologie, Über die Welt*, transl. by Hans Strohm, 3rd edition, Berlin: Akademie-Verlag 1984. Cf. the editor's extensive introduction, and Malcolm Wilson, *Structure and Method in Aristotle's Meteorologica: A More Disorderly Nature*, Cambridge: Cambridge University Press 2013.
3 Don Rittner, *A to Z of Scientists in Weather and Climate*, New York: Facts on File 2003, xiii.
4 On the early medieval and Islamic era, see Tofigh Heidarzadeh, *A History of Physical Theories of Comets, from Aristotle to Whipple*, Berlin, Heidelberg and New York: Springer Science & Business Media 2008, 23ff.; Selin, op. cit., 1666ff.
5 Cf. Thomas Aquinas, *Commentary on Aristotle's Meteorology*, transl. by Pierre Conway, O.P. and F.R. Larcher, O.P., 1964, http://dhspriory.org/thomas/english/Meteora.htm, accessed 1 November 2019.
6 Agustín Udías, *Jesuit Contribution to Science: A History*, Dordrecht: Springer 2015, 157.
7 Craig Martin, *Renaissance Meteorology: Pomponazzi to Descartes*, Baltimore, MD: John Hopkins University Press 2011, 2.
8 Wilson, op. cit., 6. Cf. Craig Martin, "The Ends of Weather: Teleology in Renaissance Meteorology," *Journal of the History of Philosophy* 48, 3, 2010, 259–282, 260.
9 Strohm, op. cit.; Kerstin Andermann and Undine Eberlein, "Einleitung: Gefühle als Atmosphären? Die Provokation der Neuen Phänomenologie," in: Kerstin Andermann and Undine Eberlein (eds.), *Gefühle als Atmosphären: Neue Phänomenologie und philosophische Emotionstheorie* (Deutsche Zeitschrift für Philosophie, Sonderband, Vol. 29), Berlin: De Gruyter Akademie Forschung 2011, 123.
10 In the 13th century, however, Aristotle's limitation of the weather to the earth and atmosphere was condemned by Étienne Tempier, Bishop of Paris, who was unnerved by the fact that the so-called secondary causes such as weather were believed to function properly even if the first cause, God, was removed from the scene. Cf. Peter Moore, *The Weather Experiment: The Pioneers Who Sought to See the Future*, London: Random House 2015, 101f.
11 Gregory of Nazianzus, *Oratio* 14:32, in a direct quote from Aristotle's *Meteorology* 1:6.
    Cf. note 123, p. 65, in: *Gregory of Nazianzus: Select Orations*, translated by Martha Vinson (The Fathers of the Church: A New Translation), New York: The Catholic University of America Press 2003.
12 Strohm, op. cit., 127.
13 Kristi McKim, *Cinema as Weather: Stylistic Screens and Atmospheric Change*, Routledge 2013, 37.

14 Ibid.
15 Cf. Martin, op. cit., 4.
16 Ibid., 264.
17 Ibid., 273, 274–278, about Luther and his followers' understanding of weather in the light of providence. Here God's will, for the Reformed theologians, represented the direct cause of meteorological events.
18 Martin, op. cit., 281f.
19 Tomek Bartzak, "Weather + Science," in: Jürgen Mayer H. and Neeraj Bhatia (eds.), *Arium: Weather + Architecture*, Ostfildern: Hatje Cantz 2010, 40–53, 40.
20 Cf. Alexander von Humboldt's essay *On Isothermal Lines, and the Distribution of Heat over the Globe*, Edinburgh 1820. Cf. Jan Munzar, "Alexander von Humboldt and His Isotherms: On the Occasion of the 150th Anniversary of the First Map of Isotherms," *Weather, Royal Meteorological Society*, 1967; Andrea Wulf, *Alexander von Humboldt und die Erfindung der Natur*, 2nd edition, München: Bertelsmann 2016 (*The Invention of Nature: Alexander von Humboldt's New World*, New York: Knopf 2015, 228–290).
21 Alexander von Humboldt, *Kosmos – Entwurf einer physischen Weltbeschreibung* (Vol. 1 1845, Vol. 2 1847, Vol. 3 1850), Frankfurt am Main: Eichborn 2004, 177 (*Cosmos: A Sketch of a Physical Description of the Universe*: www.gutenberg.org/cache/epub/14565/pg14565-images.html).
22 Humboldt, op. cit., 177.
23 Cf. Oliver Grill, *Die Wetterseiten der Literatur: Poetologische Konstellationen und meteorologische Kontexte im 19. Jahrhundert*, München: Verlag Wilhelm Fink 2019, 131f.
24 Cf. Wulf, op. cit., 228–230.
25 Ibid., 309.
26 Harald Ackermann, *Das Wetter und die Krankheiten*, Kiel 1854. Cf. Markus Breuning, *Alexander von Humboldt und die Anfänge der Medizin: Arbeiten und Kontakte Medizin und verwandte Gebiete*, Bern: disserta Verlag 2015, 78.
27 Reviewed by Goethe, op. cit., 699ff.
28 Ibid., English version first published in *London Magazine* 1821. Johann Wolfgang Goethe, *Zur Naturwissenschaft überhaupt, besonders zur Morphologie: Erfahrung, Betrachtung, Folgerung, durch Lebensereignisse verbunden*, München 1989 (1817–24).
29 Empiricism describes the elevation of the empirical method to a worldview, which claims that only the empirical is real.
30 Goethe, op. cit., II.1, 667f. Cf. (explicitly on love and compassion) *Love and Truth in Plainness Manifested, a Collection of the Several Writings of Luke Howard*, London 1704, 261.
31 John Hennig, *Goethe and the English Speaking World*, Bern: Peter Lang 1988, 201.
32 Goethe in 1815, according to A.W. Slater, "Luke Howard, F.R.S. (1772–1864) and His Relations with Goethe," *Notes and Records of the Royal Society of London* 27, 1, 1972, 119–140, 120.
33 In his short essay *Camarupa* (to Herzog Carl August in 1817), Goethe comments on the image: "Die symbolische Darstellung der Wolkenformen bringt die verschiedenen Umwandelungen wie sie vorgetragen worden, zum Anschauen". (The symbolic depictions of the cloud forms shall make visible the different conversions as these were presented.) Quoted according to *Zur Naturwissenschaft überhaupt, besonders zur Morphologie: Erfahrung, Betrachtung, Folgerung, durch Lebensereignisse verbunden*, München: Hanser 1989 (1817–24), 1061. The text was revised several times and later integrated into Goethe's poem in honour of Howard. Johann Wolfgang Goethe, *Camarupa*, 1817, in: LA (*Zur Naturwissenschaft überhaupt*) Vol. I, 11, 194–199. Cf. intriguingly Gisela Nickel, "Neues von '*Camarupa*': Zu Goethes frühen meteorologischen Arbeiten," in: Jochen Golz, Bernd Leistner, and Edith Zehm (eds.), *Goethe Jahrbuch 117*, Weimar: Verlag Hermann Böhlaus Nachfolger 2000, 118–125.

34 On Goethe's detailed examination of Howard's analysis of the clouds, see Karl-Heinz Bernhardt, "Johann Wolfgang von Goethes Beziehungen zu Luke Howard und sein Wirken auf dem Gebiet der Meteorologie," *Proceedings of the International Commission on History of Meteorology* 1, 1, 2004, 28–40, www.meteohistory.org/2004proceedings1.1/ pdfs/03bernhardt.pdf, accessed 26 April 2016. Cf. Slater, op. cit.

35 J.W. Goethe, *Howards Ehrengedächtnis*, 1817, erweitert 1821, in: *Zur Naturwissenschaft überhaupt*, 472, 612–616 (English translation, *In Honour of Howard*, 613–615).

36 Johann Wolfgang Goethe, *Bedeutende Fördernis durch ein einziges geistreiches Wort*, 1823, 306, in: *Zur Naturwissenschaft überhaupt*, 306–309.

37 Percy Bysshe Shelley, *The Cloud*, 1820.

38 See   www.mu.nl/en/about/agenda/weather-or-not-8-clouds-from-both-sides-by-karo lina-sobecka, 13 January 2020 and cf. Karolina Sobecka, "The Atmospheric Turn," in: Sigurd Bergmann and Forrest Clingerman (eds.), *Arts and Religion Responding to the Environment: Exploring Nature's Texture* (Studies in Environmental Humanities 6), Leiden: Brill Rodopi 2018, 43–58. For a photo, see also figure 8.7 in the final chapter.

39 Johannes Stückelberger, *Wolkenbilder: Deutungen des Himmels in der Moderne*, München: Verlag Wilhelm Fink 2010, 10f.

40 Cf. on this consensual evaluation in art history Sigurd Bergmann, *In the Beginning Is the Icon: A Liberative Theology of Images, Visual Arts and Culture*, London and New York: Routledge 2009, 43.

41 Stückelberger, op. cit., 367.

42 In the sense of Pierre Simon Laplace's famous comment about there being no need for the God hypothesis ("Je n'avais pas besoin de cette hypothèse-là"), allegedly as a reply to Napoleon in 1801.

43 Animism represents an essential human capacity to perceive and interact with non-human life forms as living beings with unique and individual, person-like identities which are rooted in invisible but fully experiential life forces. Without speculating, we can regard animism as an essential human skill. Following Marx's criticism of fetishism beyond traditional animism, one could ask if animism, or better neo-animism, is able to develop a countervailing power of resistance against fetishism as a central cultural force in late modern capitalism. Cf. Graham Harvey, "Introduction," in: G. Harvey (ed.), *Handbook of Animism*, London: Acumen 2014, 1–12; Sigurd Bergmann, "Fetishism Revisited: In the Animistic Lens of Eco-Pneumatology," *Journal of Reformed Theology* 6, 2012, 195–215. The challenge to revisit animism touches the heart of modernity itself and the cultural capacity of fetishisation appears as a central method in the ongoing globalisation which perverts and damages inter-subjective relations as well as human-natural relations. How this process even impacts on the commodification of weather will be discussed in the following chapter.

44 Donald Ahrens, *Meteorology Today: An Introduction to Weather, Climate, and the Environment*, 6th edition, Pacific Grove: Brooks/Cole 2000, 16.

45 Ibid., 15.

46 Franz Mauelshagen, *Klimageschichte der Neuzeit*, Darmstadt: WBG 2010, 7.

47 Andrew John Herbertson, *Outlines of Physiography: An Introduction to the Study of the Earth*, 3rd edition, London: Edward Arnold 1901, 118, https://archive.org/details/ outlinesofphysio00herb/page/118, accessed 8 November 2019.

48 Helmut Kraus, *Die Atmosphäre der Erde*, 4th edition, Berlin, Heidelberg and New York: Springer 2004, 3.

49 Ibid., 315ff.

50 Cf. Stefan Hofer and Simon Meisch (eds.), *Extremwetter: Konstellationen des Klimawandels in der Literatur der frühen Neuzeit*, Tübingen: Nomos 2018, 11 (editors' introduction).

51 According to journalists and meteorologists, terms such as "Starkregenereignis" (heavy rainfall event) and "Unwetterwarnung" (severe weather alert) have just recently been introduced into everyday language and have therein had a "steile Karriere" (bold

career). Christian Milankovic, "Meteorologen in Stuttgart: Das nächste Unwetter im Blick," *Stuttgarter Zeitung* 9 June 2016, www.stuttgarter-zeitung.de/inhalt.meteo rologen-in-stuttgart-das-naechste-unwetter-im-blick.71f5ea41-9b48-4dd3-8e2b-c0604906b724.html, accessed 14 January 2020.

52 www.dwd.de/DE/leistungen/warnwetterapp/warnwetterapp.html?nn=16124, accessed 31 October 2018.

53 Here it is important to differentiate between the first kind of prediction of weather and the second kind of prediction of climate, where the latter focuses on longer periods and applies different numeric models.

54 Marcel Proust, *In Search of Lost Time, Vol. III: The Guermantes Way*, London: Vintage Classics Edition 2000–2002, 398. Cf. Adam Watt, *The Cambridge Introduction to Marcel Proust*, Cambridge: Cambridge University Press 2011; Eve Kosofsky Sedgwick and Jonathan Goldberg (eds.), *The Weather in Proust*, Durham and London: Duke University Press 2011, 7.

55 Kraus, op. cit., 292ff.

56 Roland Barthes, *The Preparation of the Novel: Lecture Courses and Seminars at the Collège de France, 1978–1979 and 1979–1980*, New York: Columbia University Press 2011, 38. For Barthes, "*the Weather Is a Language*" (38), not simply a means of communication but also a means of instituting the weather as subject. In the frame of his wider semiotics, weather is "the code *spoken* by the moment" (38) and at the end of a day its "*communicable* essence" is "paradoxically what the weather was like"; weather is about the order of memory (39).

57 Horst Malberg, *Meteorologie und Klimatologie: Eine Einführung*, 5th edition, Berlin, Heidelberg and New York: Springer 2007, 354.

58 Mauelshagen, op. cit., 8f.

59 Ibid., 20.

60 Rüdiger Glaser, *Klimageschichte Mitteleuropas: 1200 Jahre Wetter, Klima, Katastrophen*, 3rd edition, Darmstadt: WBG 2013 (2008), 31, suggests that we should regard the approaches of thinkers in antiquity rather as a rational expansion of earlier mythological and religious weather perceptions, and acknowledges weather superstition as a vital phenomenon until the 18th century. Religious beliefs are, at least, included in his understanding of climate history, even if the concept of religion appears as all too simple and undifferentiated. The only way in which religion plays an active role in Glaser's climate history is in superstition, witch persecution, and monasteries. The field of future cooperation, however, between theology/church history/religious studies and environmental history, and especially climate history, lies hopefully and promisingly open. Glaser, op. cit.

61 Watsuji Tetsuro, *Climate and Culture: A Philosophical Study*, New York, Westport and London: Greenwood 1988 (originally published 1961).

62 Cf. Mauelshagen, op. cit.; Glaser, op. cit.; Frank Sirocko (ed.), *Wetter, Klima, Menschheitsentwicklung: Von der Eiszeit bis ins 21. Jahrhundert*, 3rd edition, Darmstadt: WBG 2012 (2009).

63 Tim Ingold, "Lines and the Weather," The Daphne Mayo Lecture, presented at the University of Queensland Art Museum on Wednesday 16 October 2013.
　Bronislaw Szerszynski, "Reading and Writing the Weather: Climate Technics and the Moment of Responsibility," *Theory, Culture and Society* 27, 2–3, 2010, 9–30.

64 Ingold, op. cit.

65 Ibid., 2.

66 Ibid., 5.

67 Ibid., 13.

68 Ibid., 14.

69 Ibid., 17.

70 Ibid., 21f.

71 Ibid., 22.

72 Ibid., 25.
73 Ibid., 28.
74 Szerszynski, op. cit., 2.
75 Ibid., 5.
76 Ibid., 15f.
77 Ibid., 19.
78 Ibid., 21. Cf. Vladimir Jankovic, *Reading the Skies: A Cultural History of English Weather, 1650–1820*, Chicago: University of Chicago Press 2001, 157.
79 Szerszynski, op. cit., 22f.
80 Ibid., 23.
81 Ibid., 27.
82 Ibid., 28.
83 Ibid., 29.
84 Szerszynski, "Life in the Open Air," in: Dirk Evers, Michael Fuller, Antje Jackelén, and Knut-Willy Sæther (eds.), *Issues in Science and Theology: What Is Life?* Heidelberg, New York, Dordrecht and London: Springer 2015, 27–42.
85 Bartzak, op. cit., 53, coins the formulation but leaves the reader not knowing what he means.
86 Cf. Henry Shue, *Climate Justice: Vulnerability and Protection*, Oxford: Oxford University Press 2014; Kimberley Thomas, R. Dean Hardy, Heather Lazrus, Michael Mendez, Ben Orlove, Isabel Rivera-Collazo, J. Timmons Roberts, Marcy Rockman, Benjamin P. Warner, and Robert Winthrop, "Explaining Differential Vulnerability to Climate Change: A Social Science Review," *Wiley Interdisciplinary Reviews Climate Change* 10, 2, 2019, www.ncbi.nlm.nih.gov/pmc/articles/PMC6472565/, accessed 30 October 2019.

For a very valuable exploration of human vulnerability, see Dorothee *Sölle, The Window of Vulnerability: A Political Spirituality*, Minneapolis: Fortress Press 1990.

6   Weather as commodity or gift?

As we saw in the previous chapter, the emergence and historical development of an empirical rational science of meteorology leads to an ambiguous situation.

On the one hand, populations are no longer helplessly exposed to weather variation as an expression of more or less comprehensible spiritual forces, but have acquired rational instruments for interpreting and understanding weather variation in the context of a universal earth system in which changes can be measured and to some degree also predicted. On the other hand, weather has lost what we might call, using Max Weber's somewhat controversial term, its "enchantment". God and the spirits have departed from the skies. Weather is no longer communicating any message but is simply there. Its variation is under some control. Vulnerability lessens. Reason replaces belief. Weather, in analogy with a great deal else in nature,[1] becomes grey.

Even if weather no longer serves as a place for spiritual revelation, nevertheless it has not lost its intrinsic power and enchantment despite meteorological explanation. Variation remains high, complex, and difficult to predict, and it can suddenly turn into extreme weather that threatens social life. Meteorology can analyse and describe what happens, but it cannot offer any assistance in coping with the existential dimension of weather's impact. Why does extreme weather hit me, here and now? Where does it come from? Why does it appear? Is God really letting the sun shine and the rain fall equally for all? Or is there some kind of injustice in the distribution of weather variation? If climate change is man-made, meteorology can only describe its consequences, and can never explain climate injustice.

On the one hand, scientific meteorology has secularised and de-sacralised weather variation; on the other hand, it has opened the way for a new mode of re-sacralising weather, that is, to perceive it as one of the last domains of nature where complexity and variation are so high that humans can respond with reverence and feel that they are at its mercy. How should one approach such a situation, in which we can partly explain weather variation meteorologically and at the same time partly become aware of our subjection to the flux of weather as an autopoetic process within a larger household of environmental forces? Should we manage weather as a commodity or accept it as a sacred gift?

This chapter will explore the commodification of weather, as it has been accelerated by the development of both modern meteorology and financial capitalism.

As weather change, and especially the emergence of extreme weather, impacts directly on vital economic processes of social life, so also meteorology is not simply an "innocent" route to empirical knowledge but deeply and actively involved in the economisation of nature and the commodification of life. In the frame of ecopolitics, too, it plays a deeply ambivalent role, where, on the one hand, it produces rich knowledge about the dynamics of anthropogenic climate change, which can serve as a central tool for social self-criticism and sustainable transformation, but, on the other hand, it establishes a forecasting industry for particular economic interests, and even opens the way for technological weather modification in the field of reengineering.

Weather and economy

The link between weather and economy is obvious but not easy to investigate. Climatic impacts are of course influencing a lot of different economic outcomes such as agriculture, health, economic growth, and security. Climate change and rising temperatures necessarily increase the need to analyse weather impacts on economically affected practices. Earlier historians held to a stereotyped view of the relation between high temperature and the skill and will to engage in physical labour. For example, Montesquieu argued in 1748 that an excess of heat made men "slothful and dispirited".[2] But the inclusion of weather in economic models is nevertheless not as developed as it could be. Dell, Jones, and Olken clearly state that weather shocks, especially shifts in temperature, affect agricultural output, industrial output, energy demand, labour productivity, health, conflict, political stability, and economic growth. Negative impacts are thus of course greater in poor countries.[3] It is still an important challenge to investigate how long-term and large-scale processes such as ongoing global warming and short-term "shocks" interact and how weather shocks influence other factors that can cause a whole series of different impacts. Moreover, the history of weather witches in the medieval period and the violent scapegoating of women for climatic change obviously had its driving force in a "a deterioration in economic conditions".[4]

The commodification of weather in meteorology mainly consists in treating "nature as information", where weather forecasts can serve as derivatives and accepted parts of business culture. In particular, the rise of insurance markets and companies has accelerated this process, as have risk capital providers and the El Niño event in 1997–1998. Moreover, energy companies have developed a stronger interest in mitigating risks with the help of meteorological analysis.[5] In view of this, Samuel Randalls strikingly talks about a "weather market" where meteorological information is instrumentalising weather so that it can serve as a commodity for trade and function within exchange processes in different environmental markets, where, for example, atmospheric pollution and global climate change put climate and weather discourse at the centre of business attention. This development has been accelerating since the 1990s, and while meteorological information was "previously relatively unvalued daily average data for many locations", it "has become a traded commodity".[6] Weather forecasts have turned into a valued commodity and so has weather itself. The atmosphere has turned from having been part of a global commons into a commodity to be bought and sold.[7]

My intention here should not be misunderstood as suggesting that belief and religious worldviews and practices could simply replace the meteorological analysis of weather. Rather, I would like to argue for a parallel strategy. On the one hand, scientific meteorology should increase its own self-critical consciousness about the risk of commodification and resist simple financial determinism in its forecasting practices. It should focus on interpreting weather as a part of the global commons where all of the world's inhabitants have a right to taste the fruits and benefit from meteorological analysis. The founding ideas of the WMO are important here, and as we saw earlier, so is the not-for-profit international cooperation on meteorological technology and information exchange between rich and poor countries. The fact that money can buy weather forecasts is nevertheless deeply problematic in the ambiguity of global monetarism, and weather (analysis) as a traded commodity should be limited in such a way that meteorology in the first place represents a practice that can benefit all citizens of the world. On the other hand, religion and beliefs must in this context not be misunderstood as magic esoteric practices that simply turn weather into a screen for self-projection, even if so-called new spiritualities offer methods for such weather wising. New Age weather wising, mostly not in accordance with traditional beliefs and practices, tends to commodify nature and weather in a way similar to the economistic worldview when it turns the inaccessibility of weather, exempt from human commands, into a screen for self-projection and doability.

Nevertheless, perceiving weather as what I would like to call a *sacred gift* must in no way take place in opposition to scientific meteorology, but it can reconstruct attitudes of reverence and grandeur in accordance with traditional and classical forms of belief. Wisdom implies deciding how to use and apply knowledge.[8] Wisdom is not necessarily equal with knowledge. Could one imagine and strive for a synthesis of meteorological knowledge with religious wisdom about the weather? Sailing a boat on the ocean, for example, demands that we listen carefully every morning and evening to the radio's meteorological weather forecast and prepare for what might come. At the same time, sailing demands an attitude where one continuously practises the belief of being at the mercy of the weather (gods), as weather can change suddenly and in unforeseen ways, whenever, wherever. A strict and exclusive consideration of only the weather forecast, without using one's own intuition and senses and without cultivating a disposition that is rooted in a spiritual attitude, appears to be foolish and even dangerous. Indeed, the meteorologist him/herself would probably agree, as the discipline includes a high degree of self-critical awareness of its own limitations with regard to the complexity and variability of its "object".

In the following, I will rewrite parts of one of the first academic essays that I ever published, entitled "Die Welt als Ware oder Haushalt?" (The World as Commodity or Household?).[9] At that time I was searching for a theme for a dissertation somewhere in the ecotheological sphere, and I was at the same time fascinated by the discovery of how much the history and culture of antiquity still might mean for a deeper understanding of our contemporary society and its views of nature. By a fortunate coincidence, a book by the famous Marxist economist and philosopher Alfred Sohn-Rethel appeared on my horizon; his usage of Werner Jaeger's classical interpretation of antiquity as "Paideia" (education, literacy, Bildung) led to a brainwave. Sohn-Rethel applied the theory of exchange to the emergence of the commodification of nature and explored the significance of the coin in this context. The coin made it possible to lend an original value to things and to regulate all processes of exchange. Commodification has impacted on human thinking and acting since then, and it has deeply influenced our understanding of nature.

In rewriting parts of this essay (which has not been published in English before), I will simply replace the notions of "nature" and "world" with "weather" and see what happens. In this way, we can explore whether the imagination and understanding of weather and our practical response to it can also be interpreted in the frame of an opposition between commodification and reverence. Surprisingly, we will see that the invention of money, which regulated exchange in antiquity, has also influenced our modern understanding of weather, even if it was the emergence of modern meteorology that first generated the conditions and accelerated its commodification.

The world/weather as commodity

From 1934 to 1947, Werner Jaeger published his extensive interpretation of Greek philosophy, entitled "Paideia",[10] in which he explored, among other things, the development of the Greek idea of cosmos.[11] According to Jaeger, the notion of "cosmos" is an idea of order and harmony that emerged in the context of the Greek city-state's development in the 6th century BC. The idea of a well organised whole where different parts are merged into an ordered formation was not transferred into social life from how the world was imagined and conceptualised but exactly the opposite: the citizens' social equality, especially in the sphere of law, which was achieved by the Greek *polis*, made possible the idea of unity, order, and equality. The notion was only later transferred from the social juridical cosmos of human society to the non-human world. Applying this to our theme of weather, the idea of weather as a somehow well-organised atmosphere would also then depend on a historical transfer in which ideas about the human social order are applied to weather orders. Our understanding of weather emerges from our understanding of our society.

Anaximander serves as a source for this interpretation, who, according to Jaeger, deduced the law-based order of nature from the juridical regulations of human society. By contrast, Xenophanes later deduced the idea of a cosmos from the idea of the world and extracted from this the model for a just and ordered human society. Later still, Heraclitus applied the cosmos idea to the human soul.

Alfred Sohn-Rethel took up Jaeger's thesis in his work on the development of the notion of "commodity" and its impacts on human thinking and acting.[12] While Jaeger points out that the view of nature in antiquity was originally shaped by social experiences, Sohn-Rethel investigates how the experiences of producing manual work and the experiences of barter impact on the view of nature in antiquity. Might the practices of manual work and barter also lie at the root of our imagination of weather?

The existence of a notion of commodity is not self-evident. Such a notion emerges first from the human practice of exchange. Significant for the emergence of a notion of commodity is the function of the coin. Sohn-Rethel departs from the observation that the pre-Socratic philosophers' search for a primal substance constituting all matter is manifested through the coin. With the help of the coin it becomes possible to lend a symbolic primary value to all material.

Experiences with the production of goods that are determined for trade in the exchange system influence the mindset and cognition of humans. From Jaeger we know that the view of society and the view of nature (and the weather) are interconnected. Sohn-Rethel substantiates this connection by claiming that experiences with exchanging traded commodities, and the mode of thinking in terms of "exchange abstraction", are transferred from the society to nature. Experiencing nature in the shape of goods exchange is generalised in such a way that nature in general is reduced to nature as commodity. Applied to weather, this means that weather, too, in all its complexity, variability and unpredictability, is reduced to a commodity that can be controlled and traded.

In the context of barter, the difference between the practice of exchange and the value of exchange is significant. The value of a commodity is abstracted from its original function. This mode of thinking in terms of "exchange abstraction" with the help of the coin represents an achievement of human reason. This corresponds to the concept of nature that frees people from experiencing nature and lends nature in the shape of a commodity with a financial, money-based value. According to Sohn-Rethel, the relation of humans and nature in antiquity is characterised by this abstract knowledge about nature in the form of a commodity. "Nature in the sense of classical mechanics can in the same way as antiquity's cosmology be described as commodified nature".[13] Sohn-Rethel entitles this process "functional socialisation" of nature[14] and identifies it in the organisational forms of Egypt, antiquity, and modern societies.

Can we also describe the modern understanding of weather as a commodity in such a sense? Even if there is much to be said for this, one cannot easily regard weather itself as a commodity, as it still totally overwhelms our human capacity and cannot be managed. Even if geoengineering clearly fits into Sohn-Rethel's theory, as it treats weather as an object to be engineered and steered according to particular human interests, one can wonder if the ongoing experiments will ever lead to any real management. Talking about weather as a commodity must therefore be limited to human practices of exploring, perceiving, and describing weather. Commodifying weather is to commodify and trade socially interpreted forms of weather.

Another part of Sohn-Rethel's analysis dealt with the structure of class society in the Greek state. Manual labour was regarded as work for slaves, and the producers of goods and riches, which allowed the equality and wealth of the citizens, were themselves excluded from the fruits of their own labour. The separation of manual and intellectual labour, of exchange and value, and of citizens and slaves also implied the separation of human and non-human life.

The interaction with nature was determined by class society. The relation of the slave, working with his/her hands, to his/her environment was of another kind than that of the citizen who was allowed to think and act freely. These two different relations were not in opposition to each other as long as the distribution of goods was unchanged. Theoretical knowledge about nature first became socially necessary for the sake of production in the modern age.

One can probably also apply Sohn-Rethel's observations about the significance of working slaves and free citizens in the social classes of antiquity and beyond to the perception of weather. Until the end of the 19th century, large parts of populations were located in the countryside and the production of food was much more labour intensive than in contemporary agribusiness times. Citizens who do not work manually and outdoors develop a different experience, perception, and understanding of their natural environment. Such a difference has been verified in many ethnographic studies and it is still true today – for example, we can trace substantial differences between tourists' short-term experience of a landscape and the experience of inhabitants who need to make their living in the landscape under different economic conditions. The question of in what way the perception of weather is influenced by identities that belong to a social class seems nevertheless to be much more complicated than is the case in Sohn-Rethel's perspective. His observations about the difference between manual and intellectual workers, slaves and free citizens can help to formulate an exciting research question, that is, how social identity and belonging impact on the perception of weather. But it can scarcely offer clear answers. Such answers might be found through a large-scale social study of how perceptions and concepts of the weather, and responses to it, are connected to different social criteria, such as education, indoor-outdoor work, gender, age, tourism, etc.

My short experiment, however, of rewriting parts of the older essay on "the world as commodity or household" with regard to weather is able to show that Sohn-Rethel's reflections on the commodification of the world can also throw some light on what we might call the commodification of weather and its early historical roots in antiquity. In modern times, it has become possible for the first time to include the analysis of weather in the social practices of exchange, and scientific meteorology has made it possible to turn the prediction of weather into forecasts which can serve as traded commodities in an exchange process that is money based and regulated by financial systems. Both business corporations and nation-states are today able to buy and trade not the weather itself but the analysis of it.

I cannot see anything problematic in producing meteorological forecasts for particular interests and social actors, be it farmers, energy institutions, or insurance companies. But as modern meteorology represents a highly expensive activity, in need of high-tech equipment such as satellites, measuring stations, and advanced computer programs, one should be careful in regarding this as an investment for profit. Observing and analysing weather change represents without doubt common goods that should remain available for the benefit of all the world's citizens. Turning forecasts into traded commodities runs a dangerous risk of privatising and particularising the tools of meteorology, which should be used and developed for the benefit of the atmosphere and the earth itself, rather than for particular money-making interests.

Weather – "guest of God"

My argument for a synthesis of scientific meteorology and spiritual reverence for weather as a sacred gift (for all) might have the capacity to produce some antidotes against such a privatisation and commodification of weather. Theologians in the early church resisted the commodification of the world, as analysed by Sohn-Rethel, and regarded the world as God's creation – that is, as a gift to the creatures rather than a product. The pre-Socratic understanding of the material as a substance was relativised, and the material was no longer eternal and a mass that could be shaped but was rather created by God from nothing. The material carried in such a view the quality of being able to reveal God's love. The bodily experiential world was in this way freed from the power of being ruled over and instrumentalised and it gained an autonomous relation to the Divine. Gregory of Nazianzus described this world as a "nature that is a guest of God" (φύσιν ξένη).[15]

The same can be said to be true for weather also. As we saw in the previous chapter, different religious traditions regarded weather not as a commodity to be controlled and traded by humans but as an all-embracing atmosphere that somehow belonged to the power of forces larger than human beings, whether it be God, gods, or spirits. A deep religious awareness of weather as a force "at the Divine's home as a guest" can also be combined with scientific meteorology. On the one hand, one does not need to believe in anything to take part in meteorological analysis and to prepare one's life according to its forecasts. On the other hand, it is fully possible to live in reverence towards life and weather, regarding them as a sacred gift, and to consider meteorology a human practice for observing and studying this gift rationally. Meteorology is then nothing more or less than looking up to the sky and encountering weather with a mix of (knowledge-based) wisdom and wonder. One could even take the spiritual quality of wondering a bit further and claim that good meteorology also needs to include a constant attitude of openness to the unpredictable and complex in its rational approaches in order to remain scientifically convincing. For Christians in late antiquity, the world as creation appeared as a divine household, as a house (oikos) in which the Creator revealed his/her love to the world. In a modern Christian perspective too, weather appears as a revelation of love, and the atmosphere serves as an all-embracing gift to humans and all other creatures. Meteorology of course by no means needs to include any religious forms of belief like this, but it can without difficulty cooperate with forms of belief, and also actively draw on these in order to avoid a fatal money-based commodification of its science. Weather affects all, even if very differently; its analysis therefore belongs to all, poor and rich, technologically developed and developing countries.

Sacred gift and global commons

My talk about weather as a "sacred gift" might seem to be confusing. A gift is always the gift of the one to the other. Who is the giver of weather? And who is receiving the gift? A gift can furthermore be rejected; weather cannot. You just have to accept it as it is. Talking about weather as a gift only makes sense in a religious context, in which the earth and life in general are perceived as God's gift to created beings. To regard the earth, including its atmosphere and weather elements, as a gift means nothing more or less than to respect the asymmetrical relation of dependence on the atmosphere and the fact that the earth can*not* be subsumed within the belief in total doability, that is, that humans are able to do everything. Human beings are dependent on the atmosphere and its different modes of weather, while the atmosphere itself is not at all dependent on us. Although human activity to some degree impacts on the function of the atmosphere, and in times of dramatic climatic change certainly also affects weather change, the relation of dependence still remains asymmetrical.

To talk about the gift as "sacred" makes sense in a religious context, as the world and weather as creation remain constantly related to the Creator, the holy one, and therefore also carry holiness in themselves. Jewish, Christian, and Muslim faiths are very clear about the difference between adoring the holy one, God the Creator, and approaching the created. Created beings and artefacts turn into idols if they are confused with the Creator God, and are therefore not to be adored. Nevertheless, all things between heaven and Earth are connected to the Divine and should therefore be approached with an attitude of belonging to the Sacred as they reveal the Creator by their own existence. The talk about the sacred is not necessarily embedded in religious language and belief but can also imply a secular meaning. "Sacred" then simply means what we regard as sacred, a selection of several phenomena that need not necessarily be connected to a religious belief in God, gods, or spirits.

In secular language one can express the same idea by talking about weather as part of the global commons. Commons cannot be commoditised. They cannot be privatised and they are "owned" by all the inhabitants of the world, and also by generations to come. Commons are, moreover, preserved regardless of their monetary value. They represent a shared right and must be passed on to future generations, at least in the same and preferably in better condition. One has no right to destroy commons. Of course, one cannot destroy the atmosphere and its weather, as these will exist and continue in their own way with or without human beings. But in my view, it also makes sense to talk about weather as a part of global commons. Human impacts on weather change and meteorological practice would then need to show consideration for the atmosphere in another way.

Climatology clearly states the impacts of anthropogenic activities on the global climate, which again affects local and regional weather change. The insight into the enormous significance of this anthropogenic impact has led to the ongoing discourse in geology and many other environmental disciplines as to whether we should identify and denote the present geological era as the "Anthropocene", the age of the humans, a discussion that we will explicitly return to in the final chapter. Regarding the atmosphere, including weather change, as global commons in this context raises serious questions about the ethical dimension of impacting on the atmosphere. Religion also changes in the era of the Anthropocene.[16] If commons should not be violated and handed over to future generations in worse condition, we must seriously object to anthropogenic activities that heavily impact on and damage the earth system. On the one hand, all human activities affect the earth, and it is not easy to draw a clear line between what kind of activities should be regarded as a violation of the commons and what should not. On the other hand, it is not difficult to analyse what kind of climatic and weather changes affect regions and people in such a dramatic life-threatening way that one can easily regard this as a consequence of abusing the right to impact on global commons, that is, the life-enhancing weather conditions for others to survive.

A second way of talking about weather as a sacred gift and global commons regards meteorology. Even if its costs are high and its technological development can only be driven by rich countries, meteorology should be regarded as an activity that must work for the benefit of all the world's inhabitants as well as non-human life systems. As weather affects everything between heaven and Earth, its scientific analysis and the limited forecast that can be deduced from it must also be available and designed for the benefit of all. If weather is regarded as a global commons, weather forecasts should also not turn into traded commodities but enrich life conditions for all.[17] Similar to health care, fresh water, and other commons, meteorology too should therefore stay within the sphere of nations and the political world community.

Meteorology for the benefit of all

Consequently, the World Meteorological Organization (WMO) is located in the institutional frame of the United Nations. It was founded in 1951, and in the aftermath of the experiences from the Second World War, the former International Meteorological Organization (IMO) was turned into a new body. This establishment seems to have been driven mainly by the interests of international scientific cooperation, but it is also interesting that meteorologists in the late 19th century were already formulating a clear political intention to let all nations benefit from its activity:

> Most important of all was the acceptance on all sides of the need to establish a permanent international body in order to ensure continued progress in the science of meteorology and also to ensure that all nations could reap the practical benefits that such progress would make possible.[18]

The intention of the new organisation was to "apply meteorology more fully to the service of mankind".[19]

Because the main interest of the scientists is in accelerating international cooperation and in the exchange and development of tools and practices, international meteorology and the WMO in itself represent a powerful instrument for peace among the nations. As weather change does not follow any territorial and national borders, the monitoring of it in meteorological science must also take place as a global and transregional activity that in itself offers a value in accordance with the vision of the United Nations. Nevertheless, meteorology can also be used for military and other power-dominating interests, and I cannot find any particular consciousness of such misuse and commodification of weather analysis in the WMO's historical documents. Nonetheless, people were already clearly aware in the 1990s about the risk of commercialisation and encouraged the organisation to ensure that "the basic principle of free and unrestricted international exchange of information among the Meteorological and Hydrological Services of Member countries" is not threatened by increasing commercial use.[20]

While I discussed earlier the risk and problems of commodifying weather (forecasting) in an ill-conceived way, in which we are still lacking an ethically well-thought-through negotiation of the modes and social contexts of forecasting, one should also point to the promising practices of global meteorology with regard to social and environmental justice. As meteorological technologies and services are incredibly expensive, these are mostly covered by the nation-states in their budgets. And as we need a global net of places where data are collected, both rich and poor countries are cooperating in a complex net of interconnected services. These services and exchange processes are not controlled and limited by financial interests but represent in themselves a marvellous practice of technology transfers between the poor and the rich. Modern global meteorology in this sense takes place as an important counterbalance to those forces of global injustice that are to be held responsible for increasing climate injustice. The advantages for all, even the planet, are obvious.[21] Poorer countries can take part in technologically advanced knowledge production and forecasting, and modern meteorology can in this way weave a stronger, geographically close-knit web, and due to this, local and regional worldwide cooperation, in combination with satellite data, develop and enhance its tools and methods. Meteorology thus also serves in itself as an instrument for equity and social and environmental justice.[22] It becomes a tool for resisting and reducing vulnerability and for increasing processes of resilience in those regions that are most affected by ongoing climate change and less capable of adapting to it.

In my view, one can clearly see in the constitutional basis of the WMO (and the ideas that originally motivated it) and its contemporary practice how an understanding of weather and climate as global commons can lead to a strong power alliance between scientists and political representatives for the people. Faith communities in different religious traditions and world regions and cultures are able to regard weather as a sacred gift that one has to approach and receive with reverence and humility. Weather remains without doubt an issue for all living beings between heaven and Earth. An alliance of believers, scientists, and political representatives might infuse hope for the future. Both as sacred gift and global commons, weather cannot turn into a commodity, and both meteorology and climate-impacting economic forces should mirror this in their practices.

While international meteorology seems to be staying on this peaceful course, actors of so-called geoengineering, which includes weather engineering, will produce more severe challenges in the future. Commercial interests are rapidly increasing and investments in expensive technical experiments are made for the sake of returning capital. At present, geoengineering experiments still seem to exist in a pioneering state where we cannot expect any immediate success.[23] As soon as an efficient form of technology is mastered, however, the situation will change, and it is therefore necessary to establish a detailed set of regulations today for an international control that is ethically and politically sound. Otherwise, it is easy to imagine the global threat that could emerge if private interests rule and apply transregionally, impacting climate technologies. But as geoengineering requires a broad and complex ethical reflection, I will not include this here but content myself with referring to the risks of technologically commoditised and engineered weather. Needless to say, a vision of weather as a sacred gift is by no means compatible with such a human engineering of weather modification.[24] Weather modification in warfare was consequently banned by the United Nations back in 1977. Could one expect a similar international consensus about a ban of sensitive technological weather modification? From our excursion to the Central Asian weather makers in the foregoing chapter, we can see that skills in impacting on weather are highly risky and that they should be strictly limited to a few specific persons who are culturally and religiously embedded.

Weaponising weather?

Before we leave this chapter open-ended, I would like to take the reader on an excursus that is frightening, at least for me. As we saw, weather manipulation as a tool in warfare was banned by the United Nations back in 1977. When I discovered this information, I merely noted it without paying any more attention to the question. Nevertheless, it was obvious that politicians of that time already had a clear awareness of the range of what they were deciding to ban.

The range of such practices of technological modification of parts of the planet's atmosphere is still difficult to imagine for me, and probably for many others. But while reading Uwe Laub's unique novel *Sturm*,[25] a strange uncomfortable and uncanny feeling spread constantly. The novel starts outside my hometown, Hannover, in an extreme hailstorm, such as one can already experience today, that damages infrastructure and also hurts the lead character's son. Worldwide, extreme weather events are increasing; many injured and dead citizens have to be taken care of in different parts of the world.

With irresistible power, the reader is sucked into a series of anxiety-producing events in which Laura, the meteorologist Daniel, and his friend and IT-specialist Leif discover how a global complot of nations (mainly China and the United States) with corporate interests are fighting each other by experimenting with a system of global weather modification in the context of a brutal reckless hunt for power over the atmosphere and the planet's inhabitants. The author is trained as a scientist and sketches skilfully a couple of convincing technologies for such modification, and his entanglement of the reader's and lead figure's naive lack of awareness of the scale of the danger and violence with detailed descriptions of military and geophysical instruments excruciatingly increases the tension of the plot. Reading the gripping story in one sitting, I was certainly chilled to the bone but nevertheless retained a feeling of safety, because this was fiction, no more no less.

Unfortunately, the meticulous author lifted this curtain and completed his work with an explanatory list of terms, projects, and institutions that blew my safety away. Looking up the terms online, I discovered how much energy national bodies have already spent on military weather modification in the past decades and how the UN ban on it has been totally ignored. While the ongoing geoengineering projects in the sphere of responding to climate change are publicly discussed and evaluated as not promising at all, the military sphere seems to operate in the cover of darkness and is still protected from a critical evaluation. This might change after Laub's brilliant novel and its wide distribution and publicity.

The Chinese Weather Modification Office became famous for claiming to have been able to impact on the weather during the 2008 Olympic Games in Beijing and to have kept the opening celebration free from rain. China has since the 1990s invested vast resources in the technology of cloud seeding, which was originally developed from military weather research in the 1940s which studied ice formation on the wings of aeroplanes.[26] This has caused strong conflicts between the country's different regions and cities, because one can talk about rain theft when some are getting rain and others are not.[27] At present, China is launching a gigantic weather "machine" intended to control rain over the dry lands of Tibet in an area the size of Alaska.[28]

Elsewhere, projects like the US "High-frequency Active Auroral Research Program" (HAARP)[29] dwelt in the grey zones of what military technological development could achieve and in fact is achieving already. For me, the similar European attempt by the EISCAT association (for the operation and development of radar facilities at high latitudes) states its intentions more openly,[30] when it claims to be working exclusively for the sake of science. According to the association's "blue book", the aim is clearly to provide facilities for "non-military scientific purposes".[31] Nevertheless, one must agree with Johan Brennan, former director of the CIA, who has aptly stated that we lack "global norms and standards" for addressing the geopolitical implications of developing such geoengineering technology.[32]

Methods of so-called cloud seeding, where one tries to bring about rain to meet agricultural needs and make financial profits as a weather maker[33] in very dry regions, seem to be innocent in comparison with experiments in the ionosphere, but these also still seem to be in a state of development such that one can wonder if they will one day lead to a reliable instrument for weather modification.[34] In the former Soviet Union, for example, scientists tried to apply the method of cloud seeding after the Chernobyl explosion, which succeeded in keeping radiation away from the region of Moscow but instead increased the density over Belarus. Although Russia denied this, it became common knowledge later.[35] Common to all such kinds of geo- and weather-engineering is the avoidance, at any cost, of an ethical discussion and evaluation of the norms, intentions, and practices. For democratic states that are already groaning under burdens of capitalist markets, such pills are hard to swallow, while dictatorships can more easily run large-scale processes even if these mostly waste resources that could be put to better use in improving conditions for poorer parts of the population.

Are power constellations like these capable of letting the demonic genie out of the bottle? And if so, can we put it back into the bottle? A new debate about the United Nations' *Environmental Modification Convention (ENMOD)* from 1977 seems undoubtedly to be necessary in order to limit and ban such further experiments with weather commodification which benefit some (whom?) and harm many other citizens of the world. The *Convention on the Prohibition of Military or Any Other Hostile Use of Environmental Modification Techniques* from 1976 also offers an international treaty that prohibits the military or other hostile use of environmental modification techniques with widespread, long-lasting, or severe effects.[36] In addition to all the suffering already caused by global warming and climate injustice, we certainly do not need any new modes of technological violation of living environments and weather conditions.

Anthropocene – is there a future beyond the era of the humans?

Finally, I would like to move our reflection on weather commodification or observance to the discourse about the Anthropocene.[37] Should we really characterise a whole historical geological period as the Era of Man/Woman? Can such a concept assist us in nurturing self-critical consciousness about the life-damaging impacts that humans have on the earth and atmosphere, including unwanted and uncontrolled weather change? Or does the notion of Anthropocene itself represent an acceptance of humans' power, an acceptance that might even accelerate processes of further fatal domination and commodification?

From the beginning, the suggestion of scholars in geoengineering and earth system analysis to talk about the Anthropocene set an optimistic tone, especially as it had grown out of a serious concern about the common future of the earth and its inhabitants. Would it become possible to break out of the fatal ideology that only technology and economy can solve the problems of a continuously accelerating environmental change with decreasing biodiversity and increasing ecological and social injustice? Might it also finally become possible to investigate the religious and "aesth/ethical"[38] driving forces of the crisis and to explore religious belief systems as both drivers and problem-solving forces?

Having followed the discussions passively for some years, my optimism has transmuted into an increasing ambivalence towards what now seems to function as a homogenising concept and a problematically generalising screen for projection. The following tries to grasp this ambiguity in three "either-ors".

The introduction of the concept of Anthropocene in 2002 bore similarities to a kind of religious process. The belief in the significance of humanity and its capacity to understand itself and, in fact, also to act as an almighty ruler over everything in the world could, in the age of the industrial revolution, be proven true by empirical science. The normative implications of such an insight remain ambivalent: they can *either* lead to a new humility towards both human and other life forms and an adequate new agenda of research questions *or* to a new triumphalist self-understanding of humankind and a utilitarian agenda regarding the human techno-economic management of using non-human (and also other human) life forms for the sake of one's own interests. Even if the introduction of the term, fortunately enough, has rather followed the humble path, one can trace among earth system scholars a certain degree of a self-aggrandising and what we might call socio-engineering attitude to the human/cultural/social/spiritual spheres of life.

For scholars in the environmental humanities in general and for scholars of religion, this ambiguity is painful. Even if the transdisciplinary potential of Anthropocene discourse is without a doubt there[39] and might encourage deeper cooperation of humanities and sciences on "the anthropogenic", the term also might serve as a catalyst for hyper-anthropocentric self-understandings in science and technology, where it becomes even more difficult to focus on the deep dependence of man/woman on nature. Will the Anthropocene narrative, then, hinder or enhance reflection on nature's complex gifts of life to the human, or what religions compress into the language of "respect for", "wisdom about" and "compassion and wonder within" nature? Might Anthropocene thinking at its worst catalyse a damaging view of nature, as in the utilitarian hyper-anthropocentric reductionism of the so-called ecosystem services, where all life in between heaven and Earth is reduced to a simple service provider for humans, a view that is radically contradictory to all religious worldviews and the fruits of environmental ethics achieved so far?

With regard to weather, we may wonder if we really should also regard the atmosphere as some kind of an "ecoservice", a view that would certainly lead to an anthropocentric commodification and economisation of the whole of weather. Will a term such as "ecoservice", which connotes that it is natural for humans to manage the earth, accelerate the striving for technological weather modification? Or can it enhance the imagination of the earth and its atmosphere as a global commons and sacred gift?

In a poignant critique of the Anthropocene narrative, Andreas Malm and Alf Hornborg have questioned the limited perspective of geologists and earth system scholars with regard to their understanding of the human species ascending to power.[40] Fittingly, they ask if the narrative "should really prompt us to abandon the fundamental concerns of social science, which importantly include the theorization of culture and power".[41] They rightly accuse the narrative of neglecting the uneven distribution of wealth as a condition for the very existence of modern, fossil-fuel technology[42] and of ignoring the fact that humans have caused global warming over the course of their long history. "Against this background", they write, " 'the Anthropocene' resembles an attempt to conceptually traverse the gap between the natural and the social – already thoroughly fused in reality – through the construction of a bridge from one side only, leading the traffic, as it were, in a direction opposite to the actual process: in climate change, social relations determine natural conditions; in Anthropocene thinking, natural scientists extend their world-views to society".[43]

For meteorology, and our understanding of its science as a work that benefits all life forms, this means that only close cooperation with environmental humanities, including the study of religion and the environment, can guarantee that meteorology will not fall into the abyss of a simple reductionist, deterministic activity to be traded as a commodity. Through such close collaboration with the humanities, it can keep its classical self-understanding alive as an investigation of the wonders of the atmosphere whose results should serve the good of all the earth's inhabitants. As we have learnt from Jaeger and Sohn-Rethel, the process of extending the view of society to the view of nature lies at the very heart of our classical tradition. Does Anthropocene thinking offer us a similar transfer, but now in the opposite direction, from the scientists' view of nature to the view of society?

One must additionally criticise the monistic understanding of humanity and its lack of contextual and historical differentiation when it talks about the human in general; humans *in general*, in my view, do not exist; humans only act in particular. With regard to the religious dimension of human beings, one might wonder if the narrative, or should we rather say trope, of the Anthropocene makes sense at all, as the history of religion shows an impressive consciousness of, and practical (for example ritual) skill in dealing with, understandings of the universal within the local, of the sacred within the profane, and of the divine and spiritual with and sometimes within the material. Other scientists in the last 200 years have also produced highly valuable notions and theories about the interaction of humanity and nature.[44] Long before the seductive trope of the Anthropocene, Alexander von Humboldt coined the inspiring notion of Earth as a *Naturgemälde* (painting of nature), where humans are both painters and painted and where nature is a lived force as well as a receiver.[45] In other words, religions and worldviews have in their history already developed advanced skills of reflecting on the reciprocal impacts of environments and human communities on each other in the light (and shadows) of the Sacred; one can only wonder to what degree earth system analysis, or meteorology, is able to include such knowledge and wisdom in its research processes or to what degree it might blur and marginalise the spiritual depth of human beings. Does a term like "faith" make any sense at all in the narrative? And should a critique of the Anthropocene narrative strike at not only its lack of reflection on the power in "uneven distribution" but also its failure to explore more deeply the ideational driving forces in the late modern *fetishism* of money and technology, as it has been analysed both in environmental anthropology and in ecotheology?[46]

For the further cultivation of the research fields of ethics and the study of religion and the environment, the ambiguity of the Anthropocene narrative seems to me to produce a problematic challenge rather than a new transdisciplinary option. But is it really an either-or? Or rather a both-and? As time goes by, I tend towards the former, but I still dare to hope for self-critical colleagues in earth system studies who might regard the Anthropocene simply as a useful trope and tool for accelerating research about the sociogenic in the anthropogenic and about the spiritual aesth/ethical in the natural, rather than for promoting a "geology of the ruling class".[47]

Nevertheless, my final and strongest objection to the Anthropocene narrative lies in the question of what we might meet *beyond* the Anthropocene. Is there space to imagine a new geological era beyond the Anthropocene? Maybe it will be the "Ecocene", where human and other life forms cohabit on Earth in fully just and peaceful entanglements? Or will it be rather an era of apocalypse where humans eradicate themselves from the planet, followed by an era of new genesis where evolution searches for new paths without human intervention? Or will it be a "post-Technocene" where the fetishisation of money and machines has been overcome and technical spaces have turned into lived spaces?[48]

Is there any thinking at all about the future in the narrative of the Anthropocene, or rather a total absence of utopia? While religions always operate to a greater or lesser degree with images of the future and so-called eschatologies, the narrative of the Anthropocene, as far as it is negotiated at present in the Anthropocene Working Group, seems to lack not only self-critical skills with regard to power, history, and ethics but also the skill to imagine a future beyond the present. How could it thereby make politically evident its social and environmental relevance? And how could one avoid the tendency of the Anthropocene discourse to both encourage and legitimate further fatal processes of commodification? Will the atmosphere preserve its grandeur, complexity, variability, alteration, and unpredictability? Or will weather become "doable" and turn into a traded commodity? Or can our awareness of the variability of weather, of its "mysteries of alteration", keep our human understanding of the future open? As we never really know what weather will bring tomorrow, its sacred gift also implies fertilising a constantly wide-open attitude to space and time, to place and future, and an antidote to human hubris. Questions like these will be deepened in the final chapter, where I argue for a new kind of atmospheric thinking on our way to the Ecocene beyond the Anthropocene. Before we climb up such onerous philosophical hills again, however, we will dwell for a while in the world of built environments, exploring how human architecture can offer shelter from different kinds of weather and how the whole practice of human drawing, building, and dwelling is embraced by weather.

Notes

1 Cf. Paul Virilio, *Grey Ecology*, New York and Dresden: Atropos Press 2009.
2 Montesquieu in *The Spirit of Laws* (1748), quoted from Melissa Dell, Benjamin F. Jones, and Benjamin A. Olken, "What Do We Learn from the Weather? The New Climate-Economy Literature," *Journal of Economic Literature 52, 3, 2014, 740–798,* 740.
3 Dell, Jones, and Olken, op. cit., 791.
4 Emily Oster, "Witchcraft, Weather and Economic Growth in Renaissance Europe," *Journal of Economic Perspectives* 18, 1, 2004, 215–228, 216.
5 For a detailed analysis, see Samuel Randalls, "Weather, Finance and Meteorology: Forecasting and Derivatives," www.meteohistory.org/2004polling_preprints/docs/abstracts/randalls_abstract.pdf, accessed 7 March 2017.
6 Ibid.

7 John E. Thornes and Samuel Randalls, "Commodifying the Atmosphere: 'Pennies from Heaven'?" *Geografiska Annaler. Series A, Physical Geography* 89, 4, 2007, 273–285.

8 Cf. Nicholas Maxwell, *From Knowledge to Wisdom: A Revolution in the Aims and Methods of Science*, Oxford: Blackwell 1984.

9 Sigurd Bergmann, "Die Welt als Ware oder Haushalt? Die Wegwahl der trinitarischen Kosmologie Gregors von Nazianz," *Evangelische Theologie* 53, 5, 1993, 460–470 (reprinted in Sigurd Bergmann, *Geist, der lebendig macht – Lavierungen zur ökologischen Befreiungstheologie*, Frankfurt am Main: Verlag für Interkulturelle Kommunikation (IKO) 1997).

10 Werner Jaeger, *Paideia: Die Formung des griechischen Menschen*, Vol. I, Berlin 1934, Vol. II, Berlin 1944, Vol. III, Berlin 1947.

11 Ibid., Vol. II, 79.

12 Alfred Sohn-Rethel, *Warenform und Denkform*, Frankfurt am Main: Suhrkamp 1971.

13 "Die Natur im Sinne der klassischen Mechanik kann ebenso wie die der antiken Kosmologie als Natur in Warenform bezeichnet werden". Op. cit., 128.

14 Ibid., 20.

15 Gregory of Nazianzus, *Oratio* 38.10.

16 Cf. Celia Deane-Drummond, Sigurd Bergmann, and Markus Vogt (eds.), *Religion in the Anthropocene: Transdisciplinary Perspectives*, Eugene, OR: Wipf & Stock/Cascade 2017.

17 Cf. Hoang, who emphasises the political dimension of the weather report and unfolds a strong and well-anchored plea for close cooperation between architects, designers, and engineers. For him, the weather report offers daily information about how to respond to climate change and weather variation if it is regarded not simply as an environmental but as a cultural project. Our next chapter will explicitly deepen Hoang's argument that "architecture and its relationship to weather is first and foremost a sociocultural project" (p. 260). Phu Hoang, "Can You Believe the Weather We're Having? The Politics of the Weather Report," in: James Graham (ed.) with Caitlin Blanchfield, Alissa Anderson, Jordan H. Carver, and Jacob Moore (eds.), *Climates: Architecture and the Planetary Imaginary*, Zürich: Lars Müller Publishers 2016, 252–260.

18 Sir Arthur Davies (ed.), *Forty Years of Progress and Achievement: A Historical Review of WMO*, Geneva: Secretariat of the World Meteorological Organization 1990, 3.

19 Ibid., 9.

20 Ibid., 179.

21 For details on this global development cooperation, see the WMO's global framework for climate services: https://gfcs.wmo.int/, accessed 31 October 2019.

22 One of many promising examples is, for example, found in the Swedish SMHI's project regarding achieving new conditions for Botswana's hydrological institutes by interconnecting hydrology and meteorology in an efficient way. www.smhi.se/nyhet sarkiv/hydrologisk-prognosverksamhet-byggs-upp-i-botswana-1.25143, accessed 31 October 2019.

23 For a convincing argument as to why one cannot rely on climate geoengineering techniques to significantly contribute to meeting the temperature goals from the Paris Agreement of the 21st UNFCCC Conference of Parties (COP21) in 2015, see Mark G. Lawrence, Stefan Schäfer, Helene Muri, Vivian Scott, Andreas Oschlies, Naomi E. Vaughan, Olivier Boucher, Hauke Schmidt, Jim Haywood, and Jürgen Scheffran, "Evaluating Climate Geoengineering Proposals in the Context of the Paris Agreement Temperature Goals," *Nature Communications* 9, 3734, 2018.

24 For a deeper discussion of geoengineering, see Christopher J. Preston (ed.), *Climate Justice and Geoengineering: Ethics and Policy in the Atmospheric Anthropocene*, London and New York: Rowman & Littlefield 2016; Mark. G. Lawrence and Paul J. Crutzen, "The Evolution of Climate Engineering Research," Opinion Article,

Geoengineering Our Climate Working Paper and Opinion Article Series 2013, http://wp.me/p2zsRk-8j; and Forrest Clingerman, "Redeeming the Climate: Investigating a Theological Model of Geoengineering," in: Celia Deane-Drummond, Sigurd Bergmann, and Bronislaw Szerszynski (eds.), *Technofutures: Transdisciplinary Perspectives on Nature and the Sacred*, Farnham: Ashgate 2015, 175–192.

25 Uwe Laub, *Sturm*, München: Heyne 2018.

26 www.futurezone.de/science/article213906021/Wettermanipulation-Das-groesste-Regenmacher-kommt-aus-China.html, accessed 4 November 2019.

27 www.heise.de/tp/features/Der-Krieg-der-Regenmacher-3435479.html, accessed 4 November 2019.

28 www.forbes.com/sites/trevornace/2018/05/10/china-is-launching-a-massive-weather-control-machine-the-size-of-alaska/#7c675d163155, accessed 4 November 2019.

29 www.gi.alaska.edu/facilities/haarp, accessed 4 November 2019. The HAARP project was closed down in 2014, after having fuelled an incredible amount of conspiracy theories: www.theweek.co.uk/world-news/58691/haarp-conspiracy-theories-what-the-mysterious-program-actually-did, accessed 4 November 2019. Laub's novel is not, of course, a documentary, but his description of the HAARP's research of the ionosphere should be taken seriously. It at least provokes the following serious question: how can large-scale technological research weaponise weather? And are such experiments going on?

30 https://eiscat.se/about/organisation/mission-statement/, accessed 4 November 2019.

31 §2a.: https://eiscat.se/wp-content/uploads/2017/06/BlueBook_Edition2015.pdf, accessed 5 November 2019.

32 https://climateandsecurity.org/2016/07/25/cia-director-on-the-geopolitical-risks-of-climate-geoengineering/, accessed 4 November 2019.

33 See for example companies like these: www.weathermodification.com/, www.australianrain.com.au/testimonials, www.weathertec-services.com/company.html, accessed 4 November 2019.

34 Cf. www.klartext-magazin.de/52A/static/wettermacher/, accessed 4 November 2019. For a detailed study of weather modification research, see the WMO's report from 2016: www.wmo.int/pages/prog/arep/wwrp/new/documents/WMO_ET_Weather_Modification_Research.pdf, accessed 4 November 2019.

35 https://sv.wikipedia.org/wiki/Molns%C3%A5dd, accessed 4 November 2019.

36 Cf. https://en.wikipedia.org/wiki/Environmental_Modification_Convention, accessed 4 November 2019.

37 The following thoughts were originally formulated in an online contribution to the Forum discussion of the Consortium for the Study of Religion, Ethics, and Society at Indiana University in Spring 2016: S. Bergmann, "Is There a Future in the Age of Humans? A Critical Eye on the Narrative of the Anthropocene," www.indiana.edu/~csres/pages/forum-folder/index.php#Bergmann, accessed 9 March 2017.

38 The concept of "aesth/ethics" brings aesthetics to the forefront of ethics. Aesthetics is here understood not as a theory of beauty in the narrow philosophical sense but as a discursive and artistic production and reflection of practices and discourses on synaesthetic perception, creation, and reception, *including* its ethical dimension. On my coinage of the programmatic term of "aesth/ethics", see Sigurd Bergmann (ed.), *Architecture, Aesth/Ethics and Religion*, Frankfurt am Main and London: IKO-Verlag für interkulturelle Kommunikation 2003; "Atmospheres of Synergy: Towards an Ecotheological Aesth/Ethics of Space," *Ecotheology: The Journal of Religion, Nature and the Environment* 11, 3, 2006, 327–357; "With-In: Towards an Aesth/Ethics of Prepositions," in: Sigurd Bergmann and Forrest Clingerman (eds.), *Exploring Nature's Texture: Engaging Environments Through Visual Arts and Religion* (Studies in Environmental Humanities 6), Leiden: Brill Rodopi 2018, 15–42.

39 Cf. Wolfgang Haber, Martin Held, and Markus Vogt in their Introduction to *Die Welt im Anthropozän: Erkundungen im Spannungsfeld zwischen Ökologie und Humanität*, München: Oekom 2016.

40 Andreas Malm and Alf Hornborg, "The Geology of Mankind? A Critique of the Anthropocene Narrative," *The Anthropocene Review* 1, 1, 2014, 62–69.

41 Ibid., 62.

42 Ibid., 64.

43 Ibid., 66.

44 Franz Mauelshagen, "A Geological Era of Mankind: The History of an Idea, and Why It Matters," in: Deane-Drummond, Bergmann, and Vogt (eds.), op. cit., 87–102.

45 On Alexander von Humboldt, see Sigurd Bergmann, "Religion at Work within Climatic Change: Eight Perceptions about Its Where and How," in: Deane-Drummond, Bergmann, and Vogt (eds.), op. cit., 68–70.

46 Cf. Alf Hornborg, "Technology as Fetish: Marx, Latour, and the Cultural Foundations of Capitalism," *Theory Culture Society*, published online 10 June 2013, http://tcs.sagepub.com/content/early/2013/06/09/0263276413488960; Sigurd Bergmann, "Fetishism Revisited: In the Animistic Lens of Eco-Pneumatology," in: *Religion, Space, and the Environment*, New Brunswick, NJ, and London: Transaction Publishers 2014, 411–428; Sigurd Bergmann, " 'Millions of Machines Are Already Roaring': Fetishised Technology Encountered by the Life-Giving Spirit," in: Deane-Drummond, Bergmann, and Szerszynski (eds.), op. cit., 115–137.

47 Daniel Cunha, "The Geology of the Ruling Class?" *The Anthropocene Review* 2, 3, 2015, 262–266.

48 For the notion of "Technocene", see Alf Hornborg, "The Political Ecology of the Technocene: Uncovering Ecologically Unequal Exchange in the World-System," in: C. Hamilton, Ch. Bonneuil, and F. Gemenne (eds.), *The Anthropocene and the Global Environmental Crisis: Rethinking Modernity in a New Epoch*, London: Routledge 2015, 57–69. For the notion of "technical space", see Sigurd Bergmann and Tore Sager (eds.), *The Ethics of Mobilities: Rethinking Place, Exclusion, Freedom and Environment*, London: Routledge 2008.

7   Under the weather roof

Shelter, faith, and architecture

Architecture as weather protection

The beginnings of human architecture are closely linked to weather. The deepest driving force to build is obviously to establish a shelter against the changing forces of weather. Buildings achieve an environment that allows the human body to dwell in balance with the changing and sometimes threatening forces of nature. Weather is therefore at the original core of architecture. "Every constructed structure that serves as protection from weather is architecture", as Heinrich Klotz strikingly puts it.[1]

The oldest forms of architecture were probably a kind of weather roof in the Stone Age, a tent made out of branches to provide shelter from rain and storms (Fig. 7.1). The origin of architecture thus lies in the need to allow the human body and community to adapt to weather conditions in the environment. And as Stone Age populations most likely lived in a spiritual universe where weather alteration was connected to the power of spiritual forces, we can in all probability also regard the oldest vernacular architecture as a human activity that is at its core connected to both weather and religion.[2]

The influential concept of the "primitive hut" (German "Urhütte") was linked to the biblical story of Adam and Eve. As the bible does not offer any description of their accommodation in paradise, one can only infer. John Milton, however, in his *Paradise Lost* describes in detail how the first hut was shaped, with a floor embroidered with flowers.[3] After their expulsion from paradise, Adam and Eve had to dwell in another hut, which Lorenzo Ghiberti depicted as a reed hut in the second panel on Cain and Abel in his *Gates of Paradise* (in the upper left) (Fig. 7.2).

Vitruvius described the hut as the beginning of architecture, where the human being creates a place in response to his/her natural environment.[4] This idea of the primitive hut strongly influenced architects and theories in the Renaissance, and especially from the 18th century onwards, and it served to push aside theories of architecture that were dominated by ideas about truth and the universal. In his influential work from 1972,[5] Joseph Rykwert discussed the constantly recurring idea that architecture has its foundation in a lost state of accord with nature. In his title, he alludes to the fabled idea that Adam had built a house in paradise, but in fact, the author is not at all interested in the perfect man's perfect house in Eden

Figure 7.1 Reconstructed Mesolithic Round-House, replica of a 10,000-year-old round-house which was excavated from a nearby clifftop site near Howick, Northumberland, Great Britain.

Figure 7.2 Lorenzo Ghiberti, *Gates of Paradise*, panel 2 (Cain and Abel), Bronze, 1425–1452, Baptistery of San Giovanni, Florence.

but in Adam's and humanity's life expelled from and outside paradise. Due to the comfortable climate, and without any need for shelter or garments, "Adam no more had a house in Paradise than Eve had a dress", Ernst Hans Gombrich writes in an expository, substantially critical review of Rykwert's book; he accuses the author of "far-flung explorations" that are not really acceptable as reliable history. For Gombrich, these appear as a "cluster of associations" and agglomeration of suggestive allusions.[6] Rykwert's method, sometimes characterised as psychoanalytical, nevertheless produces an overwhelming amount of material and insight into the broad range of narratives and images about the origin of architecture. It provides in itself an important source for the history (or historiography) of architecture, and without doubt weather and sheltering from it play an eminent role in Rykwert's stories about Vitruvius, whom he aptly called "the source of all later speculation" (Fig. 7.3).[7]

Stories about and images of the "first" house cannot of course simply be taken as historical facts but always mirror the ideas of their narrators. Architects often legitimated their own central ideas about building style with references to their (alleged) origins. Le Corbusier, for example, used a drawing of the Jewish temple in the desert to unfold his own ideas. According to him, man created a shelter for God by giving form to what was formless and used his language, geometry, to order the irregular space of nature by creating "regulating lines" (Fig. 7.4).[8]

Ideas about architecture as shelter always played a central role in these kinds of myths of the origin of architecture. Hegel expresses a similar view by emphasising architecture's function of offering a shelter for the community. In his central text about architecture, he interconnects the environmental, weather-related function with the religious. The art of architecture emerges in the context of both the Godhead's realisation and the need for defence from threatening weather:

> Architecture is in fact the first pioneer on the highway toward the adequate realization of the Godhead. In this service it is put to severe labour with objective nature, that it may disengage it by its effort from the confused growth of finitude and the distortion of contingency. By this means it levels a space for God, informs His external environment, and builds Him His temple, as a fit place for the concentration of the Spirit and for its direction towards the absolute objects of intelligent life. It raises an enclosure for the congregation of those assembled, as a defence against the threating of the tempest, against rain, the hurricane, and savage animals. It in short reveals the will thus to assemble, and although under an external relation, yet in agreement with the principles of art.[9]

According to Hegel, architecture thus both paves the way for God and contends with nature. Building as responding to the forces of weather is related to building as making a place for God.

One of my first impulses towards working on the theme of this book was provided at a visit to the German Architecture Museum in Frankfurt/M., where the

Figure 7.3 Allegorical Engraving of the Vitruvian Primitive Hut by Charles Eisen (1720–
1778), on the frontispiece of Marc-Antoine Laugier, *Essai sur l'architecture*,
2nd edition, 1755.

https://en.wikipedia.org/wiki/The_Primitive_Hut#/media/File:Essai_sur_l%27Architecture_-_Fron
tispiece.jpg, accessed 14 November 2019

upper floor offered a rather boring collection of dusty models. One of these, however, immediately attracted my attention and stimulated my imagination: a large
weather roof from the Amazon, simply built in a slightly curved half circle with
wooden pales and a roof of palm leaves (Fig. 7.5).

Such a construction could probably host hundreds of people for a lengthy
period of time in monsoon seasons and offer them preliminary shelter from the

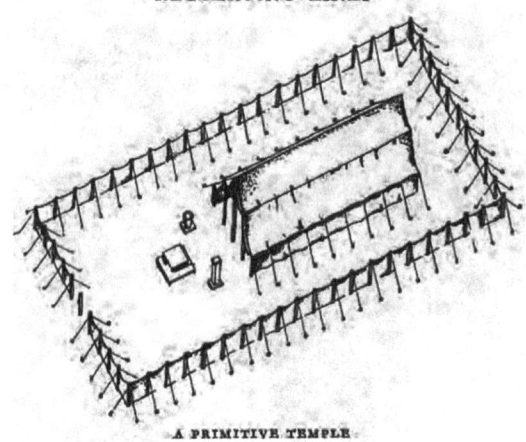

Figure 7.4 Le Corbusier, *Towards a New Architecture*, London: J. Rodker, 1931, 71.

Figure 7.5 Wetterdach, Siedlung der Yanoama-Indianer, Modell (Holz, bemalt, Pflanzen-
schaum, Sand, Jute, Moos, Staniolpapier), Atelier Ivor und Sigrid Swain, 70 ×
100 × 120 cm, 1988, Deutsches Architekturmuseum (DAM), Frankfurt/M. ©
DAM / Foto: DAM.

heavy rains. Are we encountering here the beginning of architecture? Following
Vitruvius' original myth about the primitive hut and Hegel's idea of both providing defence and paving the way for God, one can easily imagine people living
their daily lives and worshipping the spirits under such a roof (Fig. 7.6).

Figure 7.6 Traditional "atap" Roofed House, North Sumbawa, Indonesia, built January 2019.

Photo © Michael S. Northcott 2019

In what follows, this chapter will take a walk of wonder through the world of built environments where aspects of weather and faith are in communication. The foregoing chapter on culture and religion has already presented a first example of such architecture in the Fijis, and this will search for similar buildings in the history of architecture and religion. Environmental and meteorological dimensions are not necessarily significant in sacred architecture, which instead follows building methods, styles, and ideologies from its different contexts without any specific ideas or practices with regard to weather and the environment. Nevertheless, one can find buildings where both dimensions are deeply interconnected. These can serve as prototypes for what sacred architecture and faith in general pretends to offer, namely a making-oneself-at-home in the all-embracing cosmos. Building environments for sacred and non-sacred purposes in weather lands also demands a response to weather change in some way. To what degree can faith contribute to architecture and to what degree can architecture help faith to "pave the way" for God? It is self-evident that I will not follow Vitruvius' and others' interest in identifying the origin of (sacred) architecture and in following its evolution into the alleged climaxes of architectural history. Rather, I am interested in exploring how images of the Sacred and built responses to weather might interact in different cultural and geographical contexts. What does weather do to built faith? Can faith contribute to sustainable architecture in weather lands of alterations?

Shelter within nature in Aboriginal Australia

When the European settlers reached the Australian continent, the *terra incognita*, they regarded Aboriginals as a lower species for various reasons. One of these was the alleged absence of architecture and representative buildings in their life world. They mistakenly perceived the Aboriginals as nomads and failed to become aware of their subtle skills of building.

Aboriginal architecture varied from temporary windbreaks and shelters of bark to substantial round houses thatched with grass. Stone huts were also built. A variety of construction types were developed, in accordance with the demands of weather conditions as well as social organisation.[10] The shelters had rich, complex designs and were made of different materials. Many regional styles have undergone change over long periods. Aboriginal architecture is not simply a response to environmental conditions but is also "generated by distinct spatial and cognitive rules, constructs, and behaviours. Cultural symbols encoded in the physical form provide another overlay of architectural meanings".[11] Most groups had a large repertoire of shelter types, from which they made different selections according to the circumstances of weather, available material, purpose, planned length of stay, and actual length of stay. Shelters suitable for cold, dry, windy weather were developed, as well as those for shade and those for continuous rainfall.[12]

Religious symbolism is associated with these types of houses in Arnhem Land also, where the mythological activities of the Wagilag sisters influenced the construction of the forked post and ridgepole of the vaulted platform dwelling.[13] Other dwellings were used for ceremonial purposes, where people associated star constellations with the houses and camps of Ancestor Spirits.[14] Often, drawings on the inside of bark shelters illustrated stories from the Dreaming and they prefigured the later production of bark paintings. During the long wet season, members of the community had to spend many hours indoors in the shelter, and they used the time to tell and share stories about the ancestral past and to produce drawings about it on the interior bark walls.[15] Later traditions of bark painting, encouraged by the missionaries in the 1930s, most probably emerged from this practice. In this way, Aboriginal groups responded through their architecture both to environmental needs caused by changing weather conditions and to religious needs to fabricate visual and narrative meaning about their life (Fig. 7.7).

While Aboriginal architecture used to be perceived as primitive, the qualities are better understood today. Vernacular architecture, "architecture without architects",[16] offers a more direct and demanding relationship with the environment than in Western spheres where the urban dweller prefers "a safe and sealed retreat when the elements rage".[17] Weather forces, in such an architecture, are included in the building's construction, while modern architecture seals the building and excludes or filters weather changes.

Two different understandings of shelter are obviously at work here. While the Aboriginal is building the house *with* weather, the Western house is built *against* it. Both protect the body *from* damaging impacts, but while the one operates with a minimal and transparent border between in- and outdoors, the other operates

Figure 7.7 Village near Mt. Shannon, watercolour sketch by J.H. Le Keux in Sturt 1849, p. 254.

http://midja.org/sites/all/themes/midja-theme/images/shelter3.jpg, accessed 17 March 2017

with maximum control in raising a sealed wall between in- and outdoors. Hegel's aforementioned quote reflects this dichotomist view between free and wild versus controlled and civilised weather, when he talks about "defence against" the weather forces. While weather flows around and through the house in the indigenous world, and the house remains located within the natural weather land, it can only flow around the house in the modern world, and the building creates artificial weather inside its so-called weather shell. Building culture takes place within nature in the one, while it is executed in opposition to it in the other. Culture within nature stands against culture against nature.

Modern architecture in Australia proves that such a difference can be overcome. The Brambuk Living Culture Centre (Fig. 7.8) offers a synthesis of Aboriginal and Western building styles and methods, and it interconnects different symbolic references. The circles form the ground plan and allude to Aboriginal aesthetics. The curved form of the wooden roof applies techniques from boat building, the stone fireplace can be interpreted as a cosmic axis, and the curved ramp as the serpent from the dreamtime. Many influences and traditions are fused in this building, and traditional Aboriginal architecture serves not as a contrast but as an essential element of the building.[18] Nevertheless, the alleged primitivism and the stereotyped images of Aboriginal culture have been critically questioned as primitive and natural. But nonetheless, critics have also expressed their hope that the Centre can become "a gesture of reconciliation".[19]

Our short excursion to the Amazon and Aboriginal Australia can make modern architecture aware that one does not necessarily need to think and build in opposition to the environment and its weather alteration. Vernacular architecture, and especially indigenous local culture, has a large variety of skills and methods at its disposal by which houses are built in synergy with weather alteration. Macroclimatic conditions give direction to the buildings: levels of precipitation,

Figure 7.8 Brambuk Living Culture Centre, 1990, Gregory Burgess Architects.
Photo © Trevor Mein, by courtesy of the photographer

temperature variation, and the strength of prevailing winds affect the styles and forms. Both cultural traits and environmental contexts constitute the focus of vernacular building traditions.[20]

Especially in the context of globalisation, we can learn from vernacular traditions how to let architecture respond to cultural needs, including the needs for environmental and climatic shelter. Klotz' definition of architecture at the beginning of our chapter not only applies in this regard to vernacular architecture but can also serve as a foundational principle for architecture, constructing structures for weather protection. The challenge from vernacular architecture, however, is how weather is regarded and responded to (Fig. 7.9). Is it a threat to defend against, or a gift to encounter and to build with and within? Does architecture understand itself as a part of the flux of surrounding weather lands, or does it try to resist and control weather alteration?

Figure 7.9　Kazakh Yurt in the Altai, 14 October 2015. Credit: Alexandr Frolov, License: Creative Commons Attribution-Share Alike 4.0.

https://upload.wikimedia.org/wikipedia/commons/thumb/3/30/Kazah_Jurt2.jpg/1280px-Kazah_Jurt2.jpg, accessed 13 May 2020

Insights into the abundance of vernacular building traditions[21] provoke questions about the ethics of building, and about the need to respond to cultural and environmental demands and the values of the inhabitants who have to be involved.[22] Another pressing question is to what extent vernacular architecture can contribute to the demands of a future of "sustained architecture". Should one regard vernacular architecture in itself as sustained architecture, as Paul Oliver claims,[23] as it appropriates cultural and environmental needs? How can it develop in a context of rapidly growing urbanisation and population growth, where migration waves from the rural to the urban lived spaces also challenge building modes? How can traditions of building within weather lands be transformed in such a move to the postmetropolis? What kind of weather land is evolving in the postmetropolis?[24]

Before we take a closer look at contemporary architecture coping with weather and climate, we will move back in time and look for further inspiration on the interaction of weather and architecture in the European medieval period.

Medieval sacred architecture in climatic change

It is well known that the Northern hemisphere, along with other parts of the world to some degree, was affected dramatically by decreasing temperature in medieval times. Talking about the "Little Ice Age" might be an exaggeration, as it projects an image of constantly frozen lands and seas, but with regard to the scale of social and economic activities and the vulnerability of populations in their supply systems, which were much more directly dependent on environmental conditions, decreasing temperature without doubt produced serious threats and impacted dramatically on human society between 1290 and 1850. In two periods, between 1290 and the late 1400s and between about 1600 and 1800, the climate deteriorated substantially, with the coldest period being between 1645 and 1715, when average winter temperatures in Europe and North America were as much as 2°C lower than in the 20th century.

The following constitute some of this period's impacts on the landscape and the society. The Baltic Sea froze over, as did many rivers and lakes. While summers were wet and chilly, winters were long and cold. Crop failure, famine, epidemic diseases (such as the Black Death), and population decrease were ubiquitous. Large proportions of the populations were affected by starvation and poverty, and social unrest increased. Even the Vikings had to avoid ice on the sea and find new routes to sail. One can only wonder if the significant political and spiritual transformations in the history of the Reformation can also be interpreted in the light of this medieval climate change. Women could, as we saw earlier, be accused of weather witchery in the hunt for scapegoats, and unusual weather constellations that badly affected agriculture and supply could be interpreted theologically in the common frame of God's punishment of believers through environmental weather-based disasters. Climatic change in the medieval period offered without doubt an existential social, economic, cultural, and deeply religious challenge. It was widely believed that it was the transgressions and wars of kings and princes which had brought on the climate change and with which God was displeased.[25]

Despite all these challenges and impacts, Europe was also flourishing culturally and economically in the same period. The Netherlandish painters from that time offer some of the finest winter landscape paintings, which one should not simply perceive and interpret as a collection of meteorological data for climate history, as Wolfgang Behringer does a bit one-sidedly when he describes the painters' winter landscapes as a consequence of the extreme winters in the 1560s.[26] Rather, frozen winter landscapes offered new inspiring phenomena to the sensitive artists, who now could work intensively with synergies of light, temperature, and humidity (Fig. 7.10).

While European colonisation and expansion to other continents emerged and developed in this period, due to new shipbuilding technology and trade expansion, agriculture was at the same time heavily affected by the falling temperatures, which influenced the pricing of food and caused local and regional famines. While mercantilism developed, and through both its success and failure laid the ground for later capitalism, marginal and vulnerable regions went through a painful decrease in their living conditions. The gaps between the rich and the poor and social injustice thus also increased in decreasing temperatures.

Figure 7.10 Hendrick Averkamp, *A Winter Scene with Skaters near a Castle*, 1608–1609, oil on oak, 40.7 × 40.7 cm.

https://upload.wikimedia.org/wikipedia/commons/6/63/Hendrick_Averkamp_A_Winter_Scene_with_Skaters_near_a_Castle.jpg, accessed 11 November 2019

In this time period we can also observe the emergence of a new state policy in response to the urgent demands of climate-related disasters. The Danish-Norwegian state, for example, established a system of royal granaries to cope with the years of crop failure in the 1770s. These were to guarantee sufficient seeding material for the local population and also to counteract price decline in bad years. For the first time, national policies appeared for the protection of local populations who were suffering under climatic variation.[27] We can perhaps speak of these as pioneering early national climate adaption policies (Fig. 7.11).

In our context, it is highly interesting that the development of architecture and especially of sacred architecture was also affected by the decreasing temperatures

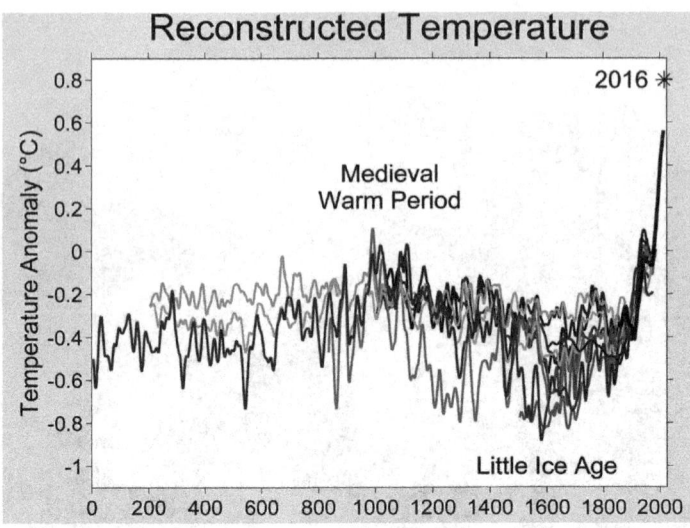

Figure 7.11 Reconstructed Temperature Charts for the Last 2000 Years, several sources combined (most recent in red, see the color on the website) with contemporary measurements.

https://commons.wikimedia.org/wiki/File:2000_Year_Temperature_Comparison.png, accessed 7 November 2019

in the Little Ice Age. Historians of architecture and archaeologists have in recent years, when cultural studies has also embarked on research into climate impact change, collected extensive empirical material on this impact, and I will here content myself with discussing only one of their findings. This leads to the question of whether the Little Ice Age's climate change can be considered a significant driving force for, if not the root of, the shift from Romanesque to Gothic architecture. In what way did dramatically changing weather conditions impact on building styles, and is it convincing that the rise of a new medieval weather land induced and accelerated the development of Gothic sacred architecture? Several strong reasons can be given for such a deep connection between weather and sacred architecture, even if, in my view, we should be careful with monistic causal explanations.

In an extensive study, incorporating a vast number of Gothic churches in different parts of Europe, Chris Simmons, a historian of architecture, has investigated whether and how the decreasing temperature and changing climate affected building methods and styles. Following studies of the geographer and historian William Wachs,[28] he focuses especially on the change in climate conditions in the Little Ice Age's periods of cooling and attempts a "more holistic climate-oriented interpretation of Medieval architecture".[29] According to Wachs, the changes that took place during this epoch in the construction techniques of sacred churches and cathedrals from the Romanesque and Gothic architectural periods (spanning nearly 600 years of European history from 950 to 1550) are clearly associated with adaptations to the climate transition. Simmons emphasises particularly the difference between the warmer medieval period before 1200, during which dry, sunny weather prevailed, and the cooler, much damper, rainy, and cloudy weather of the Little Ice Age, especially in Europe's northern regions. Romanesque architecture was adapted to the warmer, dryer conditions, with smaller windows and low-slope roofs, and it remained popular after the shift to Gothic architecture predominantly in the southern parts of France and Italy, which might be explained by the warmer climate in these regions. Architects, according to Simmons, were challenged to develop more suitable modes of building in the changing weather, and increasing precipitation, rain, and snow, and long periods of dim light due to the cloud cover, were met with high slope roofs and larger windows, characteristic of the Gothic style.

In a detailed study, Simmons also compares different regions and buildings in Europe over time and interconnects these with the change of the North Atlantic Oscillation pattern (NAO) and the movement of the jet stream over Europe southwards in this period. Not only can the substantial mode of construction be analysed with regard to changing climate patterns but the aesthetic innovations on the exterior and in the interior of Gothic churches also seem to be connected to new weather conditions. According to Simmons, roofs, paintings, sculptures, windows, domes, spires, and gargoyles "can occasionally provide a valuable illustration of changing climatic patterns".[30]

In particular, the construction of the churches' exterior reveals a careful and creative art of building where the slope of the roof, types of tiles, drainage systems, and buttresses provided new responses aimed at achieving stability in the demanding weather land. Changes in roof inclination can be clearly linked to changes in weather, particularly precipitation.[31] Steep spires were developed to dispel rain and snow efficiently, and changes in tower structure took place in this period. Pinnacles also met the need for better water shedding on the roof by shedding rain outward. Furthermore, roof tiles and gutter systems fulfilled a special function of protection in this period. Moreover, variations in campaniles and Venetian spires over the centuries following the medieval warm period provide, according to Simmons, "a particularly good illustration of these potentially climate-related architectural changes".[32]

A highly exciting part of this development lies in the increasingly widespread use of gargoyles on Gothic churches' façades and spires. According to Simmons, gargoyles are "particularly effective devices for preventing capillary attraction because they project water running down the roof top or walls of a church or cathedral into the air away from building".[33] Simmons interprets their widespread use and density as an important climate indicator. He states that churches in warmer areas have fewer gargoyles, and also relates the dates of their construction and distribution to climatic statistics. For Simmons,[34] the fact that the gargoyle method of drainage was developed at the beginning of the Little Ice Age shows that it is a product of increasing precipitation and changing weather (Fig. 7.12).

Figure 7.12 The 12 Paisley Abbey Gargoyles That Were Replaced in 1991.

https://commons.wikimedia.org/wiki/File:Paisley_Abbey_New_Gargoyles.jpg, accessed 7 November 2019

Simmons uses the same reasoning with regard to the construction of flying buttresses, and in his view the placement and increasing size of windows provides an especially clear argument for a response to darker and cloudier phases in the new medieval weather land. By increasing the size of windows and placing them lower, sunlight could be let into the building's interior, which we know carries profound theological symbolism in Gothic sacred architecture, where God meets his people in the flow and play of light around and within the building. The stained glass windows and their much more subtle transparency also fulfil such a function.[35] The best possible lighting was achieved by such means in darker, cloudy times. Moreover, mosaics and frescos were developed as a method of achieving large-scale interior decoration and of drawing on the light available in the new climate-related environment. The transition from wall frescos to stained glass to tracery took place gradually. Simmons is certainly also well aware of non-climatic considerations that account for much of the thinking surrounding the expansion of window size, such as the popular association of the lighting of interiors with the "True Light" of God,[36] and similar theological justifications of stained glass. Nevertheless, the transfer from low-transparency stained glass to windows that allowed a greater amount of interior illumination might also have some relation to the change in light conditions in the colder period. Simmons is certain that a holistic analysis provides clear evidence for increased light transmission in the North of Europe and its embedding in the Little Ice Age's climate conditions.[37]

Concluding his study, Simmons affirms a deep connection between climate and architecture style and offers a great deal of substantial evidence for the enormous consequences for architecture of the 2°C average temperature decrease in the medieval Little Ice Age. Changing weather lands without doubt impact dramatically on the architecture and affect its artistic creativity, symbolic fabrication, and technical skill in responding to changing climate conditions.

One must nevertheless critically evaluate the weight of the arguments even in such an excellent and thought-provoking study as this. Certainly, the author is committed to revealing the deep and subtle interconnections between changing weather and architectural responses. In this sense, Simmons offers an argument well-deserving of our attention – that the Gothic churches also fulfil our introductory thesis that all architecture protects from weather. But nonetheless, architecture is not *only* about sheltering from weather. Under the weather roofs other needs and functions also dwell, such as interpreting the encounter with God and his/her creation in a built environment. Power constellations need to find their form in stone, that is, in built environments. Skills of artists and craftsmen develop a dynamic of their own. The Gothic cathedrals in particular also served as nods to pilgrimage mobility, and thus had an expansive and increasing economic significance. Symbolic, spiritual, economic, cultural, artistic, engineering, and political reasons also impacted in a certain way on the development of the Gothic cathedral. Nevertheless, Simmons is convincing in his argument that the dramatic and demanding change in climate conditions has so far not received sufficient attention in the historical narrative and I hope that future research in the history of architecture will be willing to include the significance of climate change in its complex picture.

My critical review of Simmons' study does not necessarily question its merit. We should without doubt take the climate dimension seriously into account, but in my view we cannot necessarily explain everything based on it. I believe weather change instead provides an accompanying driving force that initiated and accelerated other symbolic aesthetic and political processes in the architectural development. But once in progress, other forces might have played a more significant role. The historical challenge thus implies the need to disentangle several elements and to explore how these might have interacted. If weather played an initial role, when and where did other interests and forces have their impact?

In particular, the spreading and magnificently artistic production of gargoyles seems to support such a more complex view. Once it was accepted that buildings had to be constructed in a new way in order to resist heavy rain and snowfall, the production of gargoyles might have been motivated by aesthetic and spiritual considerations rather than simple climate affordance. In featuring a complex world of demonic forces that at that time were believed to operate as aerial beings, the façade provides both an efficient drainage system and a highly inspiring and complexly narrating world of spiritual beings, often regarded as marginal art.[38]

Weather faith and architecture undoubtedly take place in a reciprocal encounter of technical, artistic, and theological forces, built in stone and animated with water flowing through grotesque faces and bodies from the sky to the earth. But heaven forbid that we establish a clear causal relation between weather and building. If bad weather, that is, cooler and wetter, led to Gothic architecture, was it then the end of the Little Ice Age and a warmer period that led to the Baroque building style?[39] Human beings are certainly bodily alive and respond to environmental and climatic change, but they are also responding to many other sociocultural forces. Art history and medieval belief can never simply provide a screen for meteorology, but one should nevertheless include the meteorological dimension in a holistic historiography.

Towards sustainable architecture

Architecture was, as we have seen, understood as a structure constructed to protect from weather both in indigenous and in medieval sacred topographies. How are architects in contemporary times responding to weather? While modern technical methods allow us to almost entirely exclude weather impacts, or better, keep them outside, the challenges of dealing with environmental and social processes are accumulating. The ongoing global urbanisation that drives larger parts of the population from the countryside to urban built environments increases risk and burdens, especially in developing countries. Air pollution, waste management, overpopulation, poverty, extreme weather events with floods and landslides, rising sea levels, and more are making multiple serious demands on planning, designing, and building. Vulnerability to climate change is much more pressing in poor regions of the South than in richer regions of the North. The global post-metropolis is characterised not only by economic inequality and injustice but also by unjustly distributed environmental risks and burdens. Climate change and new forms of weather variation intensify this inequality and extend the vulnerability even more. They present a pattern of climate injustice in which weather itself does not play the central role but in which social and economic resources and skills are unequally distributed in such a way that those world regions that are the main drivers of anthropogenic change have the best resources to cope with it. Others, in contrast, suffer most from its consequences although their rate of emissions is not comparable. Architecture also needs to encounter the challenge of geopolitical injustice in a fatal synergy with climate change, in order to find new methods of sustainable building where in particular the affected poorer world regions acquire resources and skills for creative climate adaptation.[40] Architecture is capable of redefining the problems themselves – "it educates".[41] It is in this sense able to ask questions that might not otherwise be asked, and it fills a critical void in times of crisis. As climate change for scientists is about physical atmospheric change, and for scholars in the humanities about the narratives and iconographies concerning this change, architects and designers can contribute "not by emulating the abstract logic of the system but by authoring tangible things of the here and now".[42] For example, the entrance of innumerable "plastiglomerates" in our world justifies talking about not only the Anthropocene or Technocene but also a plastisphere. One way in which we might deal with the challenge of this enormous amount of plastic is pointed out by Meredith Miller, who outlines a "post rock architecture".

While sustainable architecture might in this way contribute to creative adaptation, the challenge to contribute to mitigation is still at hand. In particular, issues of energy usage and renewable systems, but also of building materials, have thus been discussed and developed intensively. As the discourse and ongoing development of sustainable architecture today represents a highly dynamic field,[43] I will in the following confine myself to only three points. After a short historical excursion to the picturesque antecedents of environmentally and climatically conscious architecture, we will take a closer look at some examples of sustainable building *with* weather, in order to finally formulate a plea for a greater weather awareness within modern sacred architecture.

Aware of hybridised weather

In an impressive study, Jonathan Hill explores history of architecture as history of weather. Hill regards weather as "a creative architectural force alongside the designer and user",[44] and he follows the traces of environmental awareness from the Enlightenment's empiricism to our own day. Despite the fact that he constantly fuses the notions of weather, climate, nature, and environment, Hill's work offers rich inspiration and insights into the significance of weather for the history of modern architecture. Hill identifies a picturesque and romantic thread from the 18th century to the mid-20th century. The author convincingly criticises "the technocratic notion of the architect as problem solver and moderator of climatic performance",[45] such as Le Corbusier's proposal to totally isolate the inside from the outside with a neutralising wall. Hill offers an alternative model of architecture's and the environment's mutual dependence, in which the architect acknowledges weather's creative influence. For Hill it was mainly the picturesque movement and later the romantic that initiated modern architectural environmentalism, with William Kent, John Soane, and J.M.W. Turner at its centre.

The Gothic was also the subject of renewed interest in this time of the picturesque and the romantic styles, and one can only speculate as to whether the

FONTHILL ABBEY.

VIEW OF THE WEST, & NORTH FRONTS,

Figure 7.13 T. Higham and Z. Martin, *A View of the North and West Fronts of Fonthill Abbey, 1823*. From John Rutter's *Delineations of Fonthill and Its Abbey* (1823).

https://upload.wikimedia.org/wikipedia/commons/5/54/Fonthill_-_plate_11.jpg, accessed 7 November 2019

weather affordance of the Gothic architects, which we explored previously, might have contributed this interest (Fig. 7.13).

Furthermore, ruins, both natural and specifically constructed, offered an exciting mode of entangling weather and the building, as well as life, change, and death (Fig. 7.14).

Turner especially was, as we have already seen, highly conscious of the anthropogenic impacts of early industrialisation and particularly the steam engine, and he studied and depicted carefully what Hill strikingly identifies as the rise of hybridised weather: "As industries multiplied, the weather increasingly became a hybrid of the artificial and the natural".[46]

According to Hill, the picturesque gave new emphasis to the environment, acknowledging the seasons and weather's impact on human perception and understanding. Attention was paid especially to the change of season and the development of gardens, which were regarded as parts of a larger weather pattern and a means by which to engage the subjective and social weather forms.[47] A significant role was played by London's unhealthy weather; air pollution was already

Figure 7.14 James Mérigot, engraving for *A Select Collection of Views and Ruins in Rome and Its Vicinity – Recently Executed from Drawings Made upon the Spot*, London 1815.

https://upload.wikimedia.org/wikipedia/commons/0/08/A_select_collection_of_views_and_ruins_in_Rome_and_its_vicinity_-_recently_executed_from_drawings_made_upon_the_spot_%281815%29_%2814592775128%29.jpg, accessed 25 November 2019

problematic in the 16th and 17th centuries, producing a "hellish and dismal Cloud of SEA-COALE", shaking the deep classical connection between the air as the vehicle of the soul, which the character of the people depends on, and that air "of those climates they breathe".[48] Gardening became, in this time, a kind of resistance to and protection from the malignant climate of London.

Just as Turner in Chapter 6 was interpreted as the inventor of our modern weather, Hill also acknowledges the painter as offering a masterly response to the complexity of the climate and weather of his time. For Hill, Turner is not opposing nature and industry, but in his work offers an "early indication of human-induced – anthropogenic – climate change".[49] Turner depicted a hybridised weather and he thereby depicted a parallel transformation in physical nature that we increasingly experience today. Ongoing emission-driven air pollution and climate and weather change means that we can no longer experience primordial "first nature",[50] but makes it necessary for us to perceive nature and its weather as a force distinct from human *and* natural influences. Weather has turned into a hybridised force.

Hill identifies the same emphasis in Ludwig Mies van der Rohe's architectural intentions, and his Farnsworth House at the River Fox especially exposes itself to the vulnerability of climate change in continuity with Turner's deep dive into the hybridised weather of nature and industry.

Scandinavian architects also took part in the transformation from the picturesque to the modernist environmentalist; Sten Samuelson, for example, was closely following Ludwig Mies van der Rohe in his emphasis on the flow of light into and through a house (Fig. 7.15).[51]

Most notably, Sverre Fehn is to be honoured for integrating weather conditions into the purpose of a building. Here one might emphasise especially Fehn's acknowledgement of natural daylight slowly turning into darkness without using any artificial light, and his emphasis on accommodating the cycle of seasons in his buildings, which I have explored in another context (Fig. 7.16).[52]

To summarise, Hill makes us aware of the interdependence of weather and architecture as two interrelated elements of a complex system.[53] Entangling human and natural forces creates contemporary climate and weather. He pleads strikingly for a critical awareness of weather as a valuable basis for design in all stages of building. One mode of reading his committed and historically rich and excellent work is as an idealistic plea for a mode of re-designing our built environment that, due to technocratic and economic pressures in modern planning, will probably stay in lecture rooms and philosophical chambers. Another mode, however, is to take Hill's plea seriously and to demand a sustainable architecture that does not exclude but includes weather and the larger environment in its methods and ideologies to design a habitable built environment where life is enhanced, and where weather is acknowledged in all its hybridity. Undoubtedly architecture would then be able to contribute with highly valuable elements to a habitable world to live in in the Anthropocene and beyond. Might sustainable and weather-acknowledging

Figure 7.15 Sten Samuelson and Fritz Jaenecke, *Landskrona Konsthall* (Art Gallery), 1963.

https://sv.wikipedia.org/wiki/Landskrona_konsthall#/media/Fil:Landskrona_Konsthall.JPG, accessed 14 November 2019

Figure 7.16 Sverre Fehn, *A. Bødtker House*, Oslo, 1961–1965, view of the living room.

Photo: Teigen, in Christian Norberg-Schulz and Gennaro Postiglione, *Sverre Fehn: Works, Projects, Writings, 1949–1996*, New York: The Monacelli Press, 1997, 95

architecture assist us in finding a way to an Ecocene beyond the present techno-cratic Anthropocene?

Weaponised versus life-enhancing architecture

Contemporary architecture responds to the challenge not to build against but to build with weather, both in the discourse of architectural theory and in practices of building.[54] Jürgen Mayer and Neeraj Bhatia mention several themes relevant to the encounter of architecture and climate, such as science, health, war, infrastruc-ture, catastrophes, tourism, shopping, media, and materials,[55] and their section on "weather + war" in particular shows how architecture does not simply build a bet-ter sustainable world but also has dark, violent potential at its disposal, by which weather in built structures is turned into powerfully destructive tools for warfare. Weather turns into a weapon.[56]

Even if the majority of architects still follow the modernist line in regarding weather as an opponent, and in building for protection from it rather than build-ing with it, a dynamic discourse on sustainable architecture investigates climate instead as an instrument of architecture.[57] In such a view, "climate 'localises' architecture and makes it complex, multilayered and unpredictable".[58] Design-ing with climate has grown into a substantial challenge in a context in which architecture also needs to contribute to a social transformation into sustainable urban forms of life. Architecture designing with climate can contribute to stimulating the human senses. Weather can then be felt, seen, and experienced. The warmth of the sun, the cooling of the wind, the movement of the shadows through the seasons, the silence of green grounds, the smell of airborne aromas, and the taste of salty air near the coast can turn into rich qualities of built environments.[59] The atmospheres and variations within natural and built environments converge and interact. Building with weather allows the design of an urban lived space where the flux of weather and its variation can be perceived as a substantial gift of life. Building with weather does not apply the common principle of "form following function", but "function following climate".[60] The new principle would then be *form following function following climate*. Architecture starts to weatherise itself in a new way.

In a rich study of the significance of wind for urban architecture, Mareike Krautheim et al. explore how one can build with wind, a central element of weather. As human beings need to breathe, ventilation is one of the most essential dimensions of the built environment. Even if Krautheim et al. talk about "climate as architectural instrument", they are not instrumentalising wind but rather emphasising in how many different ways one can plan, draw, and build with wind. A saying ascribed to Aristotle summarises their intentions quite well:

We can't change the wind, but we can adjust the sails.[61]

Vitruvius was already aware of the different qualities of wind directions, and described west and east winds as comfortable when flowing through a city, while north and south winds should instead be kept away.[62] Contemporary architecture "integrates wind as a substantial climatological parameter into architectural design process",[63] a process that, due to technological advancements and applied geophysical and meteorological knowledge, can be developed into exciting built spaces. Energy emissions can be reduced in this way, and architecture can make its contribution to challenges of mitigation and adaptation for resisting global warming. A wonderful example of how to avoid electrical systems for ventilation in hot areas is the wind tower or wind catcher (Fig. 7.17).[64] Two openings on opposite sides are divided by a shaft and direct fresh air from the outside down to the living area. Air enters from one side, ventilates the room, and escapes from the other side.

It is interesting to note in our context that the editors of the extensive study also lend the spiritual dimension an important significance where wind in religious belief systems, mistakenly described as "mythologies",[65] appears as a multifaceted life force and divine revelation.

Krautheim et al. represent the state of knowledge today and take the discourse about climate change directly into the theory and practice of architecture, thus transforming architecture from a simple tool of protection and a system to enhance energy efficiency into a practice that is localised by the weather itself. In addition, older approaches such as Victor Olgay's *Designing with Climate* from

Figure 7.17 A Windcatcher at the Ganjali Khan Complex, Kerman, Iran.

https://commons.wikimedia.org/wiki/File:Windcatcher_at_Ganjali_Khan_Complex,_Kerman.jpg, accessed 8 November 2019

1963 increase architects' sensitivity to the climatic and ecological dimension. Nevertheless, Olgay's "bioclimatic" approach still remains within the ideology of controlling climate[66] and managing it in order to achieve "comfort zones", which are defined in a generalised way that excludes the experience of flux and variation in built weather lands, a phenomenological quality that can be found in vernacular as well as in avant-garde sustainable architecture today. Olgay, however, leads us to original insights that, as yet, very few architects are reflecting on. His detailed reflections on animals,[67] which build shelters with weather, might also inspire contemporary architecture. Furthermore, Olgay's connection between architectural regionalism[68] and the challenge to explicitly respond to the variation of climate regions and zones is worth remembering as an essential constant of designing with weather. Following the general architectural principles and ideas of Le Corbusier might have its risks, but Olgay makes us aware that Le Corbusier too was deeply conscious of "the symphony of climate" that we have not understood and that his demand for "an authentic regionalism" has its context in the challenge to design and build with climate.[69]

While Futurism at the beginning of the 19th century was, on the one hand, certainly capable of estimating the revolutionary power of new technologies, including for urbanism, it was, on the other hand, enthusiastically glorifying in its power as a tool for dominating and controlling the whole of life.[70] Weather was also included in the Futurists' fascist vision of modern life, in which the sun should be rationalised, as Konstantin Melnikov proposed in 1927.[71]

Rationalising the atmosphere is, as we know, also at the top of the agenda today for geo- and climate-engineering, and one can certainly wonder if the much-discussed notion of the Anthropocene might enhance such hubris at the same time as it encourages self-critical skills of imagining alternative sustainable futures. Designing and building with weather should in my view rather attempt to take urban life and its architectural spaces beyond the Anthropocene into a new phase of the Ecocene. Ecocene in this sense means not reducing the cultural and technical skills of human beings but interconnecting them with the life worlds of other creatures. If weather and climate can localise architecture, designing with weather might contribute to overcoming anthropogenic climate change and to transforming urban life into "meteorogenic social change". Anthropogenic impacts on the atmosphere are then turned upside-down and atmospherogenic impacts on society are taken seriously in an architecture localised by the weather.

Even if the Futurist movement flourished only in a very short time at the beginning of the century, it seems as if its central ideology still enjoys the best of health. Violent technical manipulation of nature goes on, on micro- as well as on macro-scales, including with regard to weather. One of the most brutal forms of rationalising weather is to weaponise it, for example in the water politics and construction of "dams of security", which are playing a central role in the warfare in the Middle East (Fig. 7.18).[72] Another way of technically designing built space with weather is to experiment with "de-emphasis" and offer nothing to see apart from "our dependence on vision itself" as found in the *blur building* (Fig. 7.19), where a fog-mass results from natural and man-made forces and where the visitor can move through the fog into the blue sky and even drink the building.[73]

As Futurism in its historical origin was immediately attacked and opposed by Dadaism, contemporary neo-Futurism and anthropocentric technocracy is also opposed by creative and critical forces in environmentalism, science, and technology, as well as in architecture. One of the important skills of the architect is shared with the climatologist: the creative work of up- and down-scaling. Deborah R. Coen pleads vehemently for a multiscalar approach with regard to sustainable architecture and identifies predecessors in the art history of the Habsburg state where large and small scales were integrated in a fascinating way that illustrated the interaction of all the atmospheric phenomena over the whole earth. Space and time were integrated at this time with the assistance of notions of *Alterswert* (age-value) and *Stimmung*, and Alois Riegl's concepts of ambience produced "a gaze associated with an important historical transition in the understanding of climate change".[74] The multiscalar *Anschaulichkeit* (visualisability) of weather in its impacts on natural and cultural landscapes, which was developed in the art of

Figure 7.18 The Hydroelectric Power Plant at Mosul Dam, Tigris River, Iraq.

https://commons.wikimedia.org/wiki/File:Mosul_Dam_hydro_power_plant.jpg, accessed 11 November 2019

Figure 7.19 The Blur Building, Swiss Expo 2002, Yverdon-les-Bains, Switzerland, by Charles Renfro, Ricardo Scofidio, and Elizabeth Diller.

https://upload.wikimedia.org/wikipedia/commons/a/a7/20020717_Expo_Yverdon_23.JPG, accessed 11 November 2019

that time, became an important instrument in designing with great awareness of weather. How can architecture today not just make visible the flux of weather but design and build in such a way that weather becomes tangible and experienceable for inhabitants?

As cities are taking more from the biosphere than they give back, the challenge of sustainable architecture and urbanism is at the heart of striving for alternative futures in our common future. Saskia Sassen regards buildings as instruments for advancing urban sustainability and emphasises the multiscalar capacities of the city and the diversity of ecologies.[75] Her vision of architecture "delegating back to the biosphere" sounds attractive, but in my view it still operates all too much with a distinction of nature and culture, climate and city. Interestingly, she observes that the biosphere decelerates with increasing scale.[76] The challenge for sustainable architecture then would lead to integrating anew the dimensions of time and space, speed and scale. Sassen loses weather and the climate on her way to the conclusion, but her observation seems to touch on one of the most essential challenges of building with weather, that is, to respond to the variation of the flux of weather with the skills of scaling up and down in an artistically sustainable way.

The theory and discourse about resilience, which impacts greatly on architecture as an attempt to build resilient cities, operates within a dichotomist frame of contrast between nature and culture. The vision of resilience pretends the possibility of achieving a balance. Once this balance has been achieved, consumerism and what she considers unsustainable modes of living can continue, just as long they do not threaten the carrying capacities. Resilience can work both "as a vehicle of sustainability and an agent of destitution".[77] Jesse M. Keenan therefore rightly contrasts resilience to adaptation, understood as "the transformative capacity to shift to alternative modes of consumption and production".[78] The task for the architect is, in his view, to develop consistent meanings for resilience and "make productive its counterpoint by promoting a robust capacity to adapt".[79] Building with weather is then not simply an attempt to achieve balance and to stay within given limits of resilience but to accept the superiority of weather and to adapt to its variability as a given. It does not strive to construct "hermetically sealed boxes" for an ideal climate and junk spaces, but weather architecture allows the atmospheric conditions of the exterior to "penetrate and cross-pollinate with our interior architectural atmospheres".[80]

Maybe Rachel Armstrong's vision for "designing the unknown"[81] through living buildings that grow, metabolise, and defend us like an immune system might take us a step further. Her *living architecture* uses microorganisms and other metabolic materials to invent a new technology, materiality, and way of thinking about architecture that is not simply a machine.[82] At the intersection of biology and technology, Armstrong constructs spaces that are "possessing" some of the properties of living beings. For example, gels that are computers can, in her world, compute through space and time using chemistry. Houses, for her, function and appear like alternative organ systems, sounding like a stomach gurgling rather than a combustion engine roaring.

The discourse about climate, weather, and architecture is, as we have seen so far, intense and productive. The final chapter will mine more deeply the challenge of interconnecting built atmospheres for the coming Ecocene. This chapter, however, will conclude with examples from three built environments where an attempt has been made to build with weather in a sensitive and conscious way.

Wind, temperature, light, and energy vivifying a built cog

Chalmers (Technical) University in Gothenburg built in 2011 the "Kuggen" office building as a part of Lindholmen Science Park (Fig. 7.20). The house was designed by architects Gert Wingårdh and Jonas Edblad and received awards for its sustainable design. It was inspired by Frank Gehry's *Ray and Maria Stata Center* at MIT (Fig. 7.21).

Not only the building's design and technology but also its function responds to the challenge of sustainable architecture, as the house serves as a digital library, learning commons, and meeting place for students, researchers, and the public for exploring new materials and forms of energy usage. The building's shape is not just symbolic (with regard to its location at a technical university) but emerges from aerodynamic criteria. Due to its complex façade and its curved form, the wind – which can be strong and cold from the North Sea – will flow around the building and cannot hit it directly. The air turbulence is therefore minimal.

Figure 7.20 Kuggen [the Cog], Chalmers Technical University, Göteborg, Wingårdh Arkitektkontor through Jonas Edblad, 2011.

Photo © Ewamarie Herklint, Göteborg, 2018. "Kuggen" means "cog" and alludes to the five floors that are shaped as five layered interlocking cogwheels.

Figure 7.21 Frank Gehry, *The Ray and Maria Stata Center*, at the Massachusetts Institute of Technology, 2004.

https://upload.wikimedia.org/wikipedia/en/e/eb/MIT_Strata_Center.jpg, accessed 14 November 2019

Nevertheless, the structure does not hide from the wind but leads its ventilating power around and into the building in a healthy way. The vibrant (dominantly red-purple-yellow-green-white) coloured triangles and squares underline this flow visually at the same time as they offer the visitor the illusion that the whole building is moving around, like a mechanical gearbox. The floors grow larger as you go higher, and they also grow towards the south. Thus, the upper floors can offer shade to the lower ones. The façade is composed of durable ceramic panels. Solar panels, which move along the sun's orbit, are constructed on the building's roof. They produce electricity and contribute to water heating. Each office can regulate its own temperature between 22°C and 26°C. The triangular windows, with their peak down, allow light to flow into the building, whereby the amount of glass decreases and energy can be saved.

Four different sustainable technologies have been intelligently integrated into this building: adaptive ventilation, adaptive light, interacting systems for heating and cooling, and solar panels. Designing with weather here interconnects wind, light, temperature, and energy.[83]

Thermodynamic, microclimatic, and seasonal weather wisdom at the Villa Farnese, Caprarola

Not only modern houses, using new technologies, but also older ones can provide advanced architecture that accommodates weather within and around its walls. The Italian Renaissance *Villa Farnese* was built in the 16th century in the town of Caprarola, ca. 50 km northwest of Rome (Fig. 7.22). It was commissioned by Cardinal Alessandro Farnese, the future pope Paul III (1534–1549). The architect, Giacomo Barozzi da Vignola, was inspired by Michelangelo. The interior hosts a whole iconographic program of Mannerist frescos designed to extol the family's history. The Swedish queen Kristina lived here from 1655 to 1659. The place has inspired many other estate gardens, and the casino and its gardens are today one of the homes of the Italian president. Furthermore, the villa has served as the location for movies such as the *Medici* (2016) and *Luther* (2003).

In our context, it is interesting to take a look at the fortified building's subtle interaction with different weather types in different seasons. The plan creates a rectangular pentagon with two identical suites of rooms: a winter apartment facing south and west, and a summer apartment facing north and east. The power of

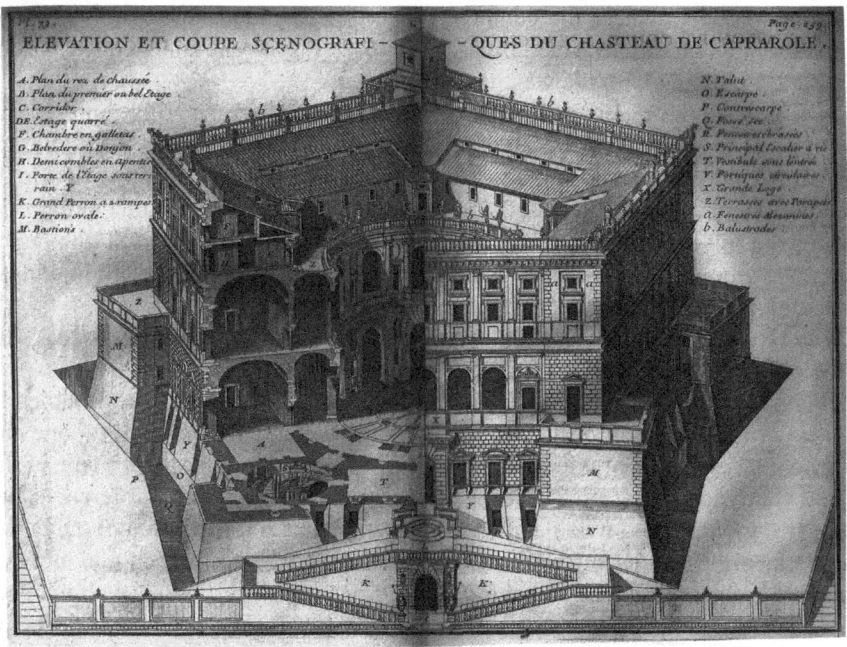

Figure 7.22 Villa Farnese (Villa Caprarola), *Prospetto principale di Palazzo di Caprarola* by Giuseppe Vasi, ca. 1746–1748.

https://commons.wikimedia.org/wiki/File:Prospetto_principale_di_Palazzo_di_Caprarola_MET_DP836552.jpg, accessed 7 November 2019

the sunlight can in this way warm the rooms in wintertime, while the strong, hot sun can be kept away in the summer season. A secret garden with seasonal plants relates to each. Rooms in the Renaissance villa are not defined by function but by position and orientation, as well as mythological themes and paths. The Villa Farnese arranges its rooms according to the cycle of the seasons. The iconographic decoration also illustrates the rhythm and change of the seasons. Manifold microclimates are included in the building and its surroundings, where terraces, gardens, pavilions, pools, and fountains occupy a variety of locations which encounter the weather in highly different ways as the seasons change. The majestic gate (moved to the Farnese Gardens on the Palatine Hills in the 19th century) symbolically presented the owner's power. The entrance gates to the lower part of the building allowed goods to be carried to the villa even in inclement weather (Fig. 7.23).

The construction of the courtyard does not only follow aesthetic principles but also offers an efficient means of cooling down the courtyard in summertime. This method has been well known in the Mediterranean region for many centuries, and the Villa Farnese enriches the method even further with a well-thought-through design of plants in the inner garden. The courtyard space can draw on different thermodynamic processes such as stratification, where hot air rises and cold air falls; convection, where the walls heated throughout the day project air upwards; and flow patterns, where the geometry of the space can produce circulating air currents. Such an efficient method also promises significant knowledge for contemporary building with the weather.[84]

Not only do the plants and gardens within and around the Villa represent an example of highly developed landscape architecture, but also the Villa's woodlands provide significant biodiversity with their many trees and other species.[85] The forest and its management of a rich and valuable diversity should therefore be regarded as an integral part of the Villa and its gardens. Together with the building's plan, the woodlands, gardens, and plantations provide an important dimension of the climatic cultivation of this place.

The weather wisdom described here does not of course only apply to the Villa Farnese but represents a skill and style of building that was common in classical Mediterranean architecture. The Italian Renaissance villa reveals how efficiently and beautifully one can design and build with weather, and how subtly one can draw on the elements of wind, temperature, light, and humidity. The shape of the round courtyard with its arcades, the biological microclimates on different scales of plants, gardens, and woodlands, the use of water, and the location of the building on the hill together offer a synergy of different methods, which undoubtedly can teach contemporary sustainable architecture a great deal. And even if not everyone needs to dwell in a similarly majestic villa, the elements of its design can be transferred to many other modes of building.

Sacred weather in the universe – St. Gabriel's Green Church

Finally, I will conclude with a sacred space where weather, architecture, and faith embrace each other. Not "protection *from* weather", as Klotz formulated it earlier,

Figure 7.23 Villa Farnese, Stairs to the Casino, probably based on designs by Giacomo del Duca, with later alterations by Girolamo Rainaldi.

www.everycastle.com/Villa-Farnese.html, accessed 19 November 2019, licensed under the Creative Commons License

but designing *with* weather represents the nucleus of architecture in my view. Such an idea and method is also at the core in St. Gabriel's Church in Toronto.[86] The Roman Catholic Parish and the Passionist Community of Canada elected in 1998 to build a new church for St. Gabriel of the Sorrowful Virgin, and decided that this should express the ecotheological ideas of the Passionist Father Thomas Berry, who is among the most influential theologians in the field of environmental philosophy and theology.[87] One may wonder if the ideas of a writer can easily be transferred into a built structure, but for the most part St. Gabriel avoids tumbling into the abyss of striving for a simple symbolic representation of theological ideas. Nevertheless, the architect succeeded in finding genuine architectural solutions that interact with ecotheology. Architecture functions in such a space rather as an interpreter of Berry's work in a common dialogue or better dia-space. The explicit intention of this church was, according to its pastor, "to establish a link between the sacredness of the gathered community of faith and the sacredness of the Earth".[88] In our climatic context especially, the building's play with light, temperature, humidity, and vegetation are of interest. God's good creation, which was so important for Berry as one "single gorgeous celebratory event",[89] embraces the believers and their liturgy even in the flux of weather that one can experience indoors as well as outdoors (Fig. 7.24).

The entire south façade of the liturgical space is glazed with clear glass in order to passively harness the winter sun's energy and to extend the sacred space of the worship area into the sacred space of the world beyond. The surrounding environment can in this way partake in the liturgy, and natural light can flow into the building. The strong power of the sun can contribute to the building's energy usage while the visitor can simultaneously feel it on his/her skin. The protruding roof offers enough shade to avoid too much warming energy in the summer.

Daylight also flows into the room through the coloured/dichroic panels that form a continuous perimeter, designed by the Welsh glass artist David Pearl. These panels are further fractured by wall-mounted dichroic-coated reflectors. The idea of letting light, and its flow in sacred space, represent the divine mystery and the spiritual quality of the cosmos has been developed for millennia. St. Gabriel relates to the Gothic cathedral's aesthetic where the path of the sun offers a movement and rhythm in time that changes the forms and tones of the colours appearing in the room and on its walls. Seasonal influences as well as the

Figure 7.24 St. Gabriel's Passionist Parish, Willowdale, Toronto, 2006, Roberto Chiotti (Larkin Architect).

Photo © Roberto Chiotti, by courtesy of the architect

character of the daily weather impact on this play, and variations of weather are transformed into bodily experiences of form and colour. Not only do the effects of the glass respond to the weather, but they also respond to the passage of time. Light, weather, and time take part in the liturgy. They are not just simple elements but living co-actors in the believers' doxology. The life-giving Spirit takes place in such a weathered creation. The emotional dimension of being bodily alive in a built environment is consciously integrated into the spatial design (Fig. 7.25).[90]

Most of the grounds of the church are devoted to the garden, and vegetation also covers parts of the walls inside the building. The garden is landscaped to recall indigenous land and wildlife habitats through the seasons.

Pedestrians can approach the church by a path through the garden with "stations of our cosmic earth", which depict significant moments in the evolutionary story of the universe and the pilgrim journey of humankind within that "story", enthusiastically interpreted by Berry and his disciples as "the universe story".[91]

A living wall reminds visitors of the importance of rainforests and their own baptismal covenants. At the same time the green wall also purifies the air of the narthex and lends the worship space a comfortable humidity.

The building is LEED™ (Leadership in Energy and Environmental Design) registered, and it uses different methods to reduce dependency on non-renewable energy resources, such as supplemental use of passive solar heating, utilising the thermal mass of the building to store and retain heat, maximising natural ventilation and daylight, use of room occupancy and daylight sensors to control electrical lighting, and use of carbon dioxide sensors to alert the building's mechanical systems when fresh air is needed.

Figure 7.25 St. Gabriel's Passionist Parish, Willowdale, Toronto, 2006, Roberto Chiotti (Larkin Architect).

Photo © Roberto Chiotti, by courtesy of the architect

The belief in the sacredness of creation, which is at the heart of Berry's writing and the discourse of ecotheology, is turned in this sacred architecture into manifold architectural practices. Sustainability is built and sacred in this place where the flux of weather vivifies the building, its surroundings, and its believers. The church becomes in this way a place of cosmic making-oneself-at-home-on-Earth.[92]

Notes

1 Heinrich Klotz, *Von der Urhütte zum Wolkenkratzer*, München: Prestel 1991, 18.
2 Paul Oliver, *Built to Meet Needs: Cultural Issues in Vernacular Architecture*, Oxford and Burlington, MA: Architectural Press 2006.
3 John Milton, *Paradise Lost*, London 1667.
4 "Thus the discovery of fire gave rise to the first assembly of mankind, to their first deliberations, and to their union in a state of society. For association with each other they were more fitted by nature than other animals, from their erect posture, which also gave them the advantage of continually viewing the stars and firmament, no less than from their being able to grasp and lift an object, and turn it about with their hands and fingers. In the assembly, therefore, which thus brought them first together, they were led to the consideration of sheltering themselves from the seasons, some by making arbours with the boughs of trees, some by excavating caves in the mountains, and others in imitation of the nests and habitations of swallows, by making dwellings of twigs interwoven and covered with mud or clay. From observation of and improvement on each others' expedients for sheltering themselves, they soon began to provide a better species of huts". Marcus Vitruvius Pollio, *De Architectura, Book II*, chapter 1:2, 30–15 BC, http://penelope.uchicago.edu/Thayer/e/roman/texts/vitruvius/2*.html, accessed 11 November 2019. Cf. Torben Dahl and Winnie Friis Møller (eds.), *Klima og arkitektur*, Copenhagen: Kunstakademiets Arkitektskoles Forlag 2008, 142.
5 Jospeh Rykwert, *On Adam's House in Paradise: The Idea of the Primitive Hut in Architectural History*, 2nd edition, Cambridge, MA: MIT Press 1981.
6 Ernst Hans Gombrich, *Reflections on the History of Art: Views and Reviews*, Berkeley and Los Angeles: University of California Press 1987, 147–151, 147.
7 Rykwert, op. cit., 5. For a slightly more balanced, though still not fully convincing, discussion of Rykwert's postulates, see Karsten Harries, *The Ethical Function of Architecture*, Cambridge, MA: MIT Press 1998, 137–140.
8 Le Corbusier, *Towards a New Architecture*, London: J. Rodker 1931, 10f.
9 Hegel, *Vorlesungen über die Ästhetik*, Einleitung, IV.3, p. 143. "Denn die Architektur bahnt der adäquaten Wirklichkeit des Gottes erst den Weg und müht sich in seinem Dienst mit der objektiven Natur ab, um sie aus dem Gestrüppe der Endlichkeit und der Mißgestalt des Zufalls herauszuarbeiten. . . ". Georg Wilhelm Friedrich Hegel, *Vorlesungen über die Aesthetik, Erster Band (Sämtliche Werke*, ed. by H. Glockner, Vol. 12), 4. Aufl. der Jubiläumsaufl. Stuttgart: Frommann 1964, 12:125.
10 For an overview, see www.australia.gov.au/about-australia/australian-story/austn-indigenous-architecture, accessed 14 March 2017, and for an extensive interpretation, see Paul Memmott, *Gunyah, Goondie + Wurley: The Aboriginal Architecture of Australia*, Brisbane: University of Queensland Press 2007.
11 Paul Memmott and Carroll Go-Sam, "19.1 Aboriginal Architecture," in: Sylvia Kleinert and Margo Neale (eds.), *The Oxford Companion to Aboriginal Art and Culture*, Oxford: Oxford University Press 2000, 405–413, 405.
12 Cf. Walter Roth, *Studies of Aboriginal Ethnoarchitectural Forms, Queensland*, 1897, https://en.wikipedia.org/wiki/Indigenous_architecture#/media/File:Queensland-aboriginal-architecture-walater-roth.jpg, accessed 7 November 2019.
13 Memmott and Go-Sam, op. cit., 407. Cf. *Repertoire of Seasonal Shelter Types Used in Eastern Arnhem Land during the Early Twentieth Century*, in: *The Oxford Companion*,

figure 235, p. 407. The narrative of the two *Wagilag Sisters* is an Aboriginal Creation story of deep cultural significance.

14 www.australia.gov.au/about-australia/australian-story/austn-indigenous-architecture, accessed 14 March 2017.

15 Cf. J. Jelínek, "A Painted Bark Shelter in Central Arnhem Land," in: Justine M. Cordwell (ed.), *The Visual Arts: Plastic and Graphic*, The Hague: Mouton 1979, 365–371.

16 Bernhard Rudofsky, *Architecture Without Architects: A Short Introduction to Non-Pedigreed Architecture*, New York: The Museum of Modern Art 1965, http://design theory.fiu.edu/readings/rudolfksy_awa.pdf, accessed 4 December 2019.

17 Memmott and Go-Sam, op. cit., 413.

18 Cf. in detail on the design www.brambuk.com.au/assets/pdf/brambukspecs.pdf, accessed 17 March 2017.

19 Kim Dovey, "19.4 Aboriginal Cultural Centres," in: *The Oxford Companion*, 419–423, 420.

20 *Encyclopedia of Vernacular Architecture of the World*, 3 volumes, ed. by Paul Oliver, of the Oxford Institute for Sustainable Development and Oxford Brookes University, Cambridge: Cambridge University Press 1997, xxiv.

21 For the dynamic discourse and research on vernacular architecture, see Oliver, op. cit., and Timothy J. Gorringe, *A Theology of the Built Environment: Justice, Empowerment, Redemption*, Cambridge: Cambridge University Press 2002.

22 Oliver, op. cit., 408.

23 Ibid., 420.

24 On the accelerating process of urbanisation that is turning the whole planet into one large city, depicted as a "postmetropolis", see Edward W. Soja, *Postmetropolis: Critical Studies of Cities and Regions*, Oxford: Blackwell 2000. Cf. Sigurd Bergmann, "Beheimatung: Making Oneself at Home with the Spirit – A Collage," chapter 1 in: *Religion, Space and the Environment*, London and New York: Routledge 2014.

25 Michael S. Northcott, *A Political Theology of Climate Change*, Grand Rapids and Cambridge: Eerdmans 2013, 41.

26 Wolfgang Behringer, *Kulturgeschichte des Klimas: Von der Eiszeit bis zur globalen Erwärmung*, München: C. H. Beck 2015, 188. Philipp Meurer has convincingly shown how climate historians often misunderstand the Netherlandish painters and use their art simply as an illustration of empirically observed weather conditions. According to Meurer, the paintings rather provide examples for a new aesthetic ability that is intrinsically tied to the existential experience of nature, life, and landscape. The popular winter scenes are to be understood as paradigmatic constructions rather than naturalistic "descriptions" of the climate. Philipp Meurer, "Representations of Reality, Constructions of Meaning: Netherlandish Winter Landscapes during the Little Ice Age and Olafur Eliasson's Glacier Series," in: Sigurd Bergmann, Irmgard Blindow, and Konrad Ott (eds.), *Aesth/Ethics in Environmental Change: Hiking through the Arts, Ecology, Religion and Ethics of the Environment* (Studies in Religion and the Environment 7), Berlin: Lit 2013, 159–176.

27 Cf. on the significance of royal granaries as a national response to climate adaptation in the 18th century Tor Ivar Hansen, "Med kongen som redningsmann: Kornmagasin som klimatilpasning på 1700-tallet," *Heimen* 52, 2015, 233–250.

28 William C. Wachs, *A Historical Geography of Medieval Church Architecture*, PhD Thesis, University of Cincinnati, Cincinnati 1964.

29 Chris Simmons, *Climate Change and Medieval Sacred Architecture*, BSc Honors Thesis, Western Illinois University 2006, 5, www.esmg.mcgill.ca/bschonorsthesis.pdf, accessed 11 November 2019.

30 Ibid., 6.

31 Ibid., 15–22.

32 Ibid., 30.

33 Ibid., 48.

34 Ibid., 51.

35 C.T. Simmons and L.A. Mysak, "Stained Glass and Climate Change: How Are They Connected?" *Atmosphere-Ocean* 50, 2, 2012, 219–240.
36 Simmons, op. cit., 92.
37 Ibid., 98ff.
38 Cf. Margrete Syrstad Andås, "Hvor marginal er marginen: Om blottere i sentrum og konger i periferien," in: Kersti Markus (ed.), *Bilder i marginalen: Nordiska studier i medeltidens konst*, Tallinn: Argo 2006, 139–158.
39 As far as I know, there is so far no explicit study of potential weather impacts on the development of Baroque art, even though Baroque artists (such as Jacob van Ruisdael) profoundly worked with weather atmospheres in their landscape paintings, and even though art history of course provides valuable insights into the artists' capturing of the atmospheric qualities of weather. See for example Sara Cornell, *Art: A History of Changing Style*, Englewood Cliffs, NJ: Prentice-Hall 1983, 260.
40 Cf. *State of the World's Cities 2008/2009: Harmonious Cities*, UN Habitat, London and Sterling: Earthscan 2008, 130–139. Poverty reduction also has an important role to play in strategies for climate adaptation.
41 James Graham et al., "Climatic Imaginaries," in: James Graham (ed.) with Caitlin Blanchfield, Alissa Anderson, Jordan H. Carver, and Jacob Moore (eds.), *Climates: Architecture and the Planetary Imaginary*, Zürich: Lars Müller Publishers 2016, 9–14, 10.
42 Meredith Miller, "Views from the Plastisphere: A Preface to Post-Rock Architecture," in: Graham (ed.), op. cit., 68–78.
43 See, for example, Graham, op. cit.; Philip Jodidio, *Green Architecture*, Köln: Taschen 2018; Mohsen Mostafavi and Gareth Doherty (eds.), *Ecological Urbanism*, Zürich: Lars Müller 2016.
44 Jonathan Hill, *Weather Architecture*, London and New York: Routledge 2012, 3.
45 Ibid., 5.
46 Ibid., 6.
47 Ibid., 84f.
48 John Evelyn, *Fumifugium, or, The Inconveniencie of the Aer and Smoak of London Dissipated together with Some Remedies Humbly Proposed by J.E. esq. to His Sacred Majestie, and to the Parliament Now Assembled*, London 1661, quoted by Hill, op. cit., 135.
49 Hill, op. cit., 174.
50 In Benjamin's sense, following his famous differentiation of first (primeval) and second (social and technical) nature. Walter Benjamin, *Selected Writings, Volume 3, 1935–1938*, Cambridge, MA and London: Harvard University Press 1996, 134.
51 Both Samuelson/Jaenecke's *Landskrona Art Gallery* and also our own villa (Löparegränden 3, Lund/Vallkärra, Samuelson/Jaenecke 1967) are most certainly influenced by buildings such as Ludwig Mies van der Rohe's *Farnsworth House*, Chicago 1951: https://en.wikipedia.org/wiki/Farnsworth_House#/media/File:Farnsworth_House_by_Mies_Van_Der_Rohe_-_exterior-8.jpg, accessed 14 November 2019.
52 Sigurd Bergmann, "Luce Nordica E Architettura Sacra: Uno Sguardo Attraverso Una Lente Teologica" (Nordic Light and Sacred Architecture – Through a Theological Lens), in: E. Bianchi, S. Bergmann, S. Calatrava Valls, D. Coutagne, J. A. Felix de Carvalho, B. Daelemans, A. Dall'Asta, D Forconi, A. Gerhards, G. Gresleri, A. Lameri, Ph. Markiewicz, Á J. de Melo Siza Vieira, J.-P. Sonnet, M. Struck, and P. Toamtis, *Architetture Della Luce: arte spazi, liturgia*, Communitá di Bose, Magnano: Edizioni Qiqajon 2016, 143–161.
53 Hill, op. cit., 321.
54 One need not interpret my suggestion of a "with-or-against" necessarily as an either-or, since building-with-weather also includes technical solutions. Banham talks, for example, about architecture's mediating between different weather conditions and describes different forms in his vision of a "well-tempered environment". Reyner Banham, *The Architecture of the Well-Tempered Environment*, London: Architectural Press 1969. Cf. Phu Hoang, "Can You Believe the Weather We're Having?" in: Graham, op. cit., 252–259, 259.

55 Jürgen Mayer H. and Neeraj Bhatia (eds.), *Arium: Weather + Architecture*, Ostfildern: Hatje Cantz 2010.

56 Johanna Bollozos, "Weather + War," in: Mayer and Bhatia (eds.), op. cit., 68–79.

57 One of the pioneers of environmentally conscious architecture, Siegfried Ebeling, approaches space as membrane. In his book *Der Raum als Membran* from 1926 (Dessau: Dünnhaupt), Ebeling investigates the relations between humans' physical and psychological capacities in the environment and emphasises especially the climatic dimension. Influenced by psychiatrist Willy Hellpach, Ebeling creates a kind of utopian bio-technological approach that was not very successful in his own time but anticipated the ecological interest in the second half of the century. Hellpach had shaped the notion of "geopsyche", describing the fact that the expression of our identity, personal and communal, is related to our environment, and that our psychology is social, spatial, and climatic. Willy Hellpach, *Die Geopsychischen Erscheinungen: Die Menschenseele unter dem Einfluß von Wetter und Klima, Boden und Landschaft*, Leipzig: Engelmann 1911. Cf. Pep Avilés, "On Membranes, Masks, and Siegfried Ebeling's Environmental *Raumkubus*," in: Graham (ed.), op. cit., 319–327.

58 Mareike Krautheim, Ralf Pasel, Sven Pfeiffer, and Joachim Schultz-Granberg, *City and Wind: Climate as an Architectural Instrument*, Berlin: DOM Publishers 2014, 6.

59 Krautheim et al., op. cit., wonderfully display how wind blows through and around houses and how one can consciously design with wind.

60 Jürgen Mayer H, "Weatherize!" in: Mayer and Bathia (eds.), op. cit., 226–228.

61 Krautheim et al., op. cit., 9.

62 Marcus Vitruvius Pollio, *De Architectura, Book I, Chapter 6*, 30–15 BC, http://penelope.uchicago.edu/Thayer/e/roman/texts/vitruvius/1*.html, accessed 12 November 2019.

63 Krautheim et al., op. cit., 13.

64 An example of such a *Badgir* in Iran is presented in Krautheim et al., op. cit., 164f. Cf. Mehdi N. Bahadori, Alireza Dehghani-sanij, and Ali Sayigh, *Wind Towers: Architecture, Climate and Sustainability*, Heidelberg, New York, Dordrecht and London: Springer 2014.

65 Krautheim et al., op. cit., 58–61.

66 Indoor climate control and air conditioning have today become globally standardised. Landing in any international airport in the world or working in a modern office will offer you the same temperature and humidity. Eva Horn, "Air Conditioning: Taming the Climate as a Dream of Civilization," in: Graham (ed.), op. cit., 233–241, 240.

67 Victor Olgay, *Design with Climate: Bioclimatic Approach to Architectural Regionalism*, 1963, 1–2 (New and expanded edition: Princeton, NJ: Princeton University Press 2015).

68 Ibid., chapter 3.

69 According to Olgay, op. cit., 13. Cf. Le Corbusier, "Building an Entire New City in India, Chandigarh," *The Architectural Forum*, 1953, 142–149.

70 On Futurism, see Sigurd Bergmann, "'Millions of Machines Are Already Roaring': Fetishised Technology Encountered by the Life-Giving Spirit," in: Sigurd Bergman, Celia Deane-Drummond, and Bron Szerszynski (eds.), *Technofutures, Nature and the Sacred*, Farnham: Ashgate 2015, 115–137.

71 Graham, op. cit., 12. S. Frederick Starr, *Melnikov: Solo Architect in a Mass Society*, Princeton, NJ: Princeton University Press 1978, 179.

72 See Zeynep S. Akinci and Pelin Tan, "Waterdams as Dispossession: Ecology, Security, Colonization," in: Graham (ed.), op. cit., 142–148.

73 https://dsrny.com/project/blur-building, accessed 11 November 2019.

74 Deborah R. Coen, "Seeing Planetary Change, Down to the Smallest Wildflower," in: Graham (ed.), op. cit., 34–39, 39.

75 Saskia Sassen, "A Third Space: Neither Fully Urban nor Fully of the Biosphere," in: Graham (ed.), op. cit., 172–179. Cf. Saskia Sassen, "Cities and the Biosphere," in: Ian Spellerberg and Daniel E. Vasey (eds.), *The Berkshire Encyclopedia of Sustainability:*

Vol. 10: The Future of Sustainability, Berkshire Publishing 2012, www.saskiasassen. com/PDFs/publications/cities-and-the-biosphere.pdf, accessed 12 November 2019.

76 Sassen, "A Third Space," op. cit., 179.

77 Jianguo Wu and Tong Wu, "Ecological Resilience as a Foundation for Urban Design and Sustainability," in: Steward Pickett, M.L. Cadenasso, and Brian McGrath (eds.), *Resilience in Ecology and Urban Design: Linking Theory and Practice for Sustainable Cities*, Dordrecht: Springer 2012, 211–229, 224.

78 Jesse M. Keenan, "The Resilience Problem: Part 1," in: Graham (ed.), op. cit., 159–162, 161f.

79 Ibid., 162.

80 Neeraj Bhatia, "The Post-Junkspace Globe: Towards Weather in Architecture," in: Mayer and Bhatia (eds.), op. cit., 236–248, 248.

81 Rachel Armstrong, *Experimental Architecture: Designing the Unknown*, London and New York: Routledge 2019. Armstrong's architecture in, with, and within lived space (in E. Soja's sense) provides the liberative opposite to the Futurist's fascist vision of "millions of machines roaring". Cf. Sigurd Bergmann, "'Millions of Machines Are Already Roaring'," in: Olgay (ed.), op. cit. www.routledge.com/Technofutures-Nature-and-the-Sacred-Transdisciplinary.

82 According to an interview in www.itsnicethat.com/articles/rachel-armstrong-univer sity-of-the-underground-architecture-040817, accessed 4 December 2019.

83 www.youtube.com/watch?v=tU7g2yzW4nI. Cf. Klassen, who in his installations uses a thermal camera to make façades visible to us in a new way and pleads for an architecture that enables us to perceive climate change and to see buildings as hot, cold, wet, dry, bright, dark, and malleable, emphasising their changing weather conditions. F. Klassen, "Convergence of Architectural, Visual and Climate Data," in: V. Echarri and C.A. Brebbia (eds.), *Eco-Architecture VI: Harmonisation between Architecture and Nature*, Southampton: WIT Press 2016, 225–236.

84 Cf. Juan M. Rojas, Carmen Galán-Marín, and Enrique D. Fernández-Nieto, "Parametric Study of Thermodynamics in the Mediterranean Courtyard as a Tool for the Design of Eco-Efficient Buildings," *Energies* 2012, *5*, 7, 2381–2403.

85 M. Agrimi, A. Borgna, R. Cantone, L. Portoghesi, and M. Romagnoli, "The Management of Woodlands within the Historic Parks: The Case-Study of Forest Stands in Villa Farnese at Caprarola (Viterbo, Italy)," in: Angela Ferrari (ed.), *Proceedings of CULTURAL HERITAGE Cairo 2009* (4th International Congress "Science and Technology for the Safeguard of Cultural Heritage of the Mediterranean Basin", Cairo, Egypt), Rome 2010, Vol. I, 40–46, https://archive.org/stream/bub_gb_7zKJOYkW0GsC/bub_ gb_7zKJOYkW0GsC_djvu.txt, accessed 12 November 2019.

86 For a discussion of theological and engineering ideas behind the building's design, see the Parish's website: http://stgabrielsparish.ca/who-we-are/green-church/, accessed 10 May 2017. For a video with architect and ecotheologian Roberto Chiotti, see www. youtube.com/watch?v=G-Z0HvgorPc, accessed 10 May 2017.

   Other examples of sustainable sacred architecture may be found in *Faith & Form: The Interfaith Journal on Religion, Art and Architecture* XLI, 1, 2008 (Education Issue: Greening God's House), and Michael J. Crosbie, *Houses of God: Religious Architecture for a New Millennium*, Melbourne: Images Publishing 2019.

87 See Mary Evelyn Tucker, John Grim, and Andrew Angyal, *Thomas Berry: A Biography*, New York: Columbia University Press 2019.

88 Roberto Chiotti, "The Architecture of Eco-Theology," *Faith & Form: The Interfaith Journal on Religion, Art and Architecture* XLI, 1, 2008 (Education Issue: Greening God's House), 6–11, 6.

89 Thomas Berry, *The Dream of the Earth*, San Francisco: Sierra Club, 1988 (reprint Berkeley: Counterpoint 2015), 5.

90 Cf. Décosterd and Rahm, who are (according to Dahl, op. cit., 150f.) "defining emotional climate data" where neurologic insights are related to architecture and where

bodily experience is included in the design of houses. Cf. Décosterd and Rahm, *Physiologische Architektur/architettura fisiologica*, Basel: Bundesamt für Kultur, Birkhäuser 2002.

91 Brian Swimme and Thomas Berry, *The Universe Story: From the Primordial Flaring Forth to the Ecozoic Era – A Celebration of the Unfolding of the Cosmos*, San Francisco: Harper San Francisco 1992. For a critical perspective on turning the whole of evolution into one "story", see Lisa Sideris, "Science as Sacred Myth? Ecospirituality in the Anthropocene Age," *Journal for the Study of Religion, Nature and Culture* 9, 2, 2015, 136–153.

92 Sigurd Bergmann, *Religion, Space & the Environment*, London and New York: Routledge 2014, chapter 1. Cf. on my aesth/ethics of nature Zoë Lehmann Imfeld and Andreas Losch (eds.), *Our Common Cosmos: Exploring the Future of Theology, Human Culture and Space Sciences*, London: Bloomsbury Publishing/T&T Clark 2018, 118f., and B. Daelemans, S. J., *Spiritus Loci: A Theological Method for Contemporary Church Architecture*, Amsterdam: Brill 2015, 39–41.

8 Atmosphere and Anthropocene
Critical considerations of a narrative and image in transit to the Ecocene

"Dreadful hot weather" in hothouse Earth

What dreadful hot weather we have! It keeps one in a continual state of inelegance,[1] novelist Jane Austen complained, in her letter to Cassandra from 18 September 1796, in the wake of an Indian summer; prolonged hot weather at a time lacking in regular washing, soap, and antiperspirants could plague people with "overpowering body odours".[2]

"Inelegance" is perhaps the least of the evils of hot weather, despite the writer's ironic exaggeration in a time when "horses sweat, gentlemen perspire, and ladies feel the heat", but for many, heat can be distressing and even dangerous. A speculative but empirically well-grounded future scenario for a Germany affected by climatic change in 2040 prognosticates that beside droughts, floods, and extreme storms, hot weather also represents a serious threat to health, especially for inhabitants of large urban areas, where the poorest will be victimised most by overheated milieus exposed to extreme solar radiation.[3] The World Medical Association of scholars in climate-related medicine lists an overwhelmingly large number of different diagnoses, explicitly limited to what harm overly hot surroundings are doing to bodily and mental health.[4]

Austen's lament, therefore, might aptly summarise the populace's common feeling in the extremely hot, dry summer of 2018 in Central and Northern Europe. While exceptional, it continued a series of hot summer periods in the past decades in line with the global warming trend. At least in this experience, the discourse about climate change moved, in the shape of bodily experienced weather, directly into citizens' ordinary life world. Climate change from now on is no longer simply abstract but bodily, and socially, emotionally, and ideationally present for all. It is present here and now and not only for locally affected people somewhere else or in limited regions such as flooded river banks or storm-ravaged settlements. Climate change, already for decades dreadfully life-endangering for many peoples in the South and for limited places and regions in the North, in the heat wave from April to September 2018 also reached the majority of those who are responsible for most of the factors in global warming, that is, the proprietary classes of the North. Climate change from now on irreversibly appears as a new troublesome, unpredictable alteration of weather. What geologists circumscribe as a "great transformation" in climatic changes in the earth system (and many other impacts of human activities such as population growth, use of resources including water, loss of biodiversity, etc.) in the time of the so-called Anthropocene is now becoming tangible, visible, and bodily perceivable in ordinary life worlds. Climate change appears as dreadfully bad weather – bad for humans, naturally, rather than for the atmosphere itself.

What happened? In 2018, the spring and summer months in Central and Northern Europe were characterised by so-called weather anomalies (below-average precipitation and above-average temperatures and hours of sunshine). From April to August large parts of the continent suffered from heat waves[5] and droughts with the largest consequences so far, with extensive forest fires,[6] crop damage, and heat damage. Safe water supply was impacted locally and regionally. Due to low water levels in large rivers, power plants had to be shut down, and inland navigation had to be limited or abandoned. Train travel was impacted by rail deformations, and UV-radiation approached dangerous levels for people exposed to the sun. Several countries faced crop losses of between 30% and 50%, on such a scale that the European Union had to adjust its subsidy payment system to support farmers with overwhelming economic problems. Furthermore, loss of hay led to scarcity of animal feed, which again forced many to slaughter animals in large numbers, which in turn led to an increase in the price of meat in Europe in winter 2018–2019. Energy prices, however, increased radically due to shortfalls in energy supply systems during the long, hot, dry summer. There is good news, however, for producers and consumers of wine, because grapes (including in the author's own garden) were of an incredible quality; the year 2018 is thus a historic one not only for meteorologists but also for vintners.

In the Northern parts of Sweden, heavy forest fires started in July 2018 and more than 25,000 hectares of forest were blazing across the country. Fire fighters from several other countries – Denmark, Norway, Finland, Italy, Germany, France, Austria, Poland, and Portugal – went to the rescue, but it took until 2019 before all the fires could finally be slaked. Southern Norway and Portugal were also suffering significantly from longer forest fires at this time, in Portugal across an area of as much as 27,000 hectares, and in Germany several fires had to be fought in dry areas such as moors or military exercise areas. Fires set by arsonists blazed out of control in Greece due to the extreme hot weather conditions. Insect populations were dramatically reduced and birds suffered from food and water shortages, while climate-change-based species migration allowed tropical ticks to move north and other vermin to increase. Recently planted trees died, and in Germany up to 85% of all saplings dried out. Fish also died on a large scale in environments where water temperature increased too much and oxygen was lacking.

The close relation of this extremely hot summer weather to ongoing climate change is evident. How can one understand, and explain, the one in terms of the other?

Meteorologically, the rise of temperatures on average from April to July 2018 reveals an extreme that supersedes all foregoing summer periods since regular records began. This period, it should be noted, followed directly upon another weather anomaly, namely the harsh late winter in the Nordic countries and in

Germany from February to early April 2018. At that time I drove with my son and my best friend between Lund in Southern Sweden and Trondheim in Mid-Norway; as we went, one week we experienced a marvellously cold and snowy weather land as we passed the mountains on our way north, and the next week we witnessed, wide-eyed, a green encounter with a kind of warmly flourishing and awakening life, as spring-summer arrived with the temperature rising about 1°C for every ten miles we drove downhill southwards from the high mountains around Hamar towards Oslo and the Swedish west coast. While writing and reporting personally on this trip, it feels as if it is rather (the uncannily alternating) *Weather itself that reports me*, in accordance with Roni Horn's upside-down perspective on who reports whom on what, a promising method for emphasising in depth the reciprocal entanglement of weather and human life.[7]

April 2018 was already the warmest month ever in both Scandinavia and Central Europe, a record that endured through May–July, as we can see in the statistics for Potsdam, Germany, in Figure 8.1.

The average temperature in this period is a surprising 4.3°C more than in the previous 30 years, and distinguished oceanographer Stefan Rahmstorf describes this measurement as "mit Abstand der grösste Ausreisser nach oben relativ zur Klimakurve" (by far the largest upward outlier, relative to the climate curve). Paul Becker, vice president for *Deutscher Wetterdienst*, summarises this aptly:

> We are experiencing this summer in many places in the world simultaneously a cumulative appearance of heat waves and extreme heavy precipitation. Exactly such an accumulation of extreme meteorological events has been predicted for us as a consequence of anthropogenic climate change.[8]

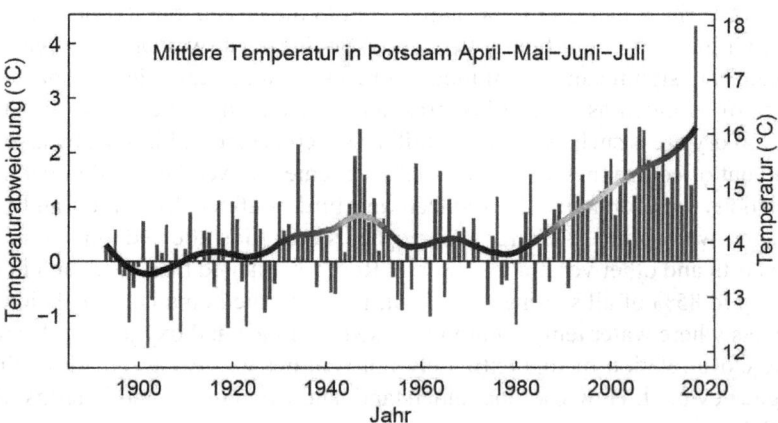

Figure 8.1 Average Temperature (right) and Deviation (left) in Potsdam April–May–June–July. Graph: Stefan Rahmstorf, Creative Commons License CC BY-SA 3.0, in Stefan Rahmstorf, *Hitze ohne Ende*, Blog: Klimalounge, 6 August 2018.

https://scilogs.spektrum.de/klimalounge/hitze-ohne-ende/

According to initiated climate impact scientists and meteorologists, the extremely hot summer of 2018 needs to be understood in the context of climate change impacting on the change in weather. Extreme weather shows us "the face of climate change".[9] In this year's case, however, the change, surprisingly enough, appears as *non-change*, that is, the lack of ordinary regular variations between low- and high-pressure areas, "negotiating" with each other as to how the summer's weather evolves. Rather, the extremes depend on a weather situation characterised by a series of blockages.

Due to the proportionally much more rapid increase in temperature in the Arctic, the jet stream that usually decides on the negotiation of low and high pressures is unstable and weakened so that the pattern of change that we are used to can no longer take place with the same frequency. As Coumou was already able to show in 2005, the jet stream has in recent decades been slowed down substantially due to the increasing temperature in the Arctic.[10] Recently, for the first time, Screen and Simmonds were able to present clear statistical evidence for the link between high-amplitude Rossby waves and surface extremes in the mid-latitudes.[11]

The jet stream, which depends on the temperature difference between the tropical and arctic spheres, is simply slowing down due to the smaller difference. Such a process means that the weather variation in the summer of the so-called mid-latitudes is less intense.[12] Weather simply becomes more persistent and stable. A specific pattern of waves, the so-called Rossby waves in high amplitudes, functions within the jet stream as a driving force, but in the kind of situation where blockages happen the reduced waves between North and South become fixed and form a specific pattern that causes the blockage where one pressure area blocks off the other. Climate scientists have been predicting for some time that global warming caused by anthropogenic emissions could lead to such blockages.

Coumou et al. investigate the impact of accelerated warming in the Arctic context, and show clearly

> that interactions between Arctic teleconnections and other remote and regional feedback processes could lead to more persistent hot-dry extremes in the mid-latitudes. The exact nature of these nonlinear interactions is not well quantified but they provide potential high-impact risks for society.[13]

One additional impact on this process was analysed by Rahmstorf, who observes how the Gulf Stream is also weakened in this situation. By measuring colder surface water on the North Atlantic south of Greenland, he was able to observe impacts on the emergence of pressure build-ups which, again, are impacting on the interrelation between low- and high-pressure areas.[14] The weakened power of the Rossby waves and of their eastward and westward travellings is furthermore highly likely to be capable of causing floods, as for example Petoukhov et al. have been able to demonstrate.[15] Accordingly, we can understand both the immense flood along the Elbe River in 2013 and the heat wave in 2018 in the same context of global warming related reduction of the high amplitude, quasi-stationary Rossby waves. In that sense, we should see different regional and continental periods of extreme weather as different threatening faces of ongoing climate change. Obviously, the human fingerprint leaves a significant mark on the atmosphere's seasonal cycles.[16]

Without doubt, therefore, the interconnection between heat waves and droughts, as well as extreme precipitation in other areas, in the situation of weakened waves and blocks caused by high pressure must be regarded as embedded in ongoing anthropogenic climate change. Absurdly, the acceleration of such dangerous and dramatic climate change can lead not just to an increased rate and expansion of change but rather to its opposite, that is, to an absence of life-enhancing weather alteration in a way that we, our co-creatures, and ecosystems are used to in an ordinary summer. Seasonal change as experienced in Europe in 2018 reveals the threatening face of climatic change, and it seems impossible to imagine that there might be any path back to common and accustomed patterns of weather variation in a future that is becoming all the more uncertain, unpredictable, and demanding. What dreadful weather that will keep us continuously in a state of inelegance, discomfort, uncertainty, trouble, and even fear!

The challenge to rapidly catalyse creative adaptation to such new atmospheric variation, in a weather land that we have not yet seen, is significant. Might we also, in a new common future, be able to join in the Creator's appreciation of his/her work? "God saw all that he had made, and it was very good" (Genesis 1:31). What about belief in a good creation on this planet Earth, where the new normal state due to human activities is at indisputable "risk of heading towards irreversible 'Hothouse Earth' state",[17] as internationally leading scientists Will Steffen, Johan Rockström, Katherine Richardson, Hans Joachim Schellnhuber et al. conclude in their widely received study. Their shocking conclusion builds on a complex exploration of trajectories and "self-reinforcing feedbacks" in the Anthropocene when Earth is "crossing a threshold", a process that one could feel in all its inelegance in one's body and soul in the remarkable summer of 2018, and likewise in several periods of the year 2019.[18] How can one believe in a continuing creation that was, is, and will be "very good" when "humanity [is] at risk for pushing the system across a planetary threshold and irreversibly down a Hothouse Earth pathway"?[19]

In remembering the small transformation from what we expected as normal into an extreme summer in 2018, we have slowly begun to search for explanations about what happened that year and in others in the past within the Anthropocene's great transformation. In the following, I will depart from the intensely dynamic discourse about the Anthropocene and explore what kind of roles weather and religion might play herein, before finally trying to bring different concepts of the atmosphere into constructive communication. As the Anthropocene concept appears upon closer examination to be a problematically depoliticising concept, we will need to seek a way to overcome its limits. Rethinking weather needs to approach the atmosphere not only as a geophysical entity but also as a multilayered sociocultural, spiritual, aesthetic, ethical, and historical agency. Connecting faith and weather in such a context will bring to the fore the fatal limits in the ongoing Anthropocene discourse and open a path beyond these where the Ecocene appears as an encouraging vision beyond contemporary violence and suffering.

Anthropocene – the great transformation

Ever since the Stratigraphy Commission of the Geological Society of London made a case in 2008 for incorporating the Anthropocene into the geological time-scale, the debate about the understanding of the Anthropocene has made massive waves.[20] The notion "Anthropocene" implies a shift from the Holocene to a new epoch in the earth's history, where human impacts since the so-called Industrial Revolution in the 18th century have increased on such a scale that

> [t]he human imprint on the global environment has now become so large and active that it rivals some of the great forces of Nature in its impact on the functioning of the Earth system.[21]

Among many other types of evidence, such as accelerating rates of species invasion and extinction (Fig. 8.2), rising sea levels, and human disturbance of the climate system (Fig. 8.3), the traces of human activities, such as nuclear waste, plastic waste, and soot, on planet Earth have increased on a significant scale.

Scientists have coined the notion of the "great acceleration" to summarise the radical shift that took place when CO_2 emissions increased and the process of global warming accelerated. The great acceleration refers to the most recent period of the Anthropocene during which the rate of impact of human activity upon the earth's geology and ecosystems is increasing significantly (Fig. 8.4).

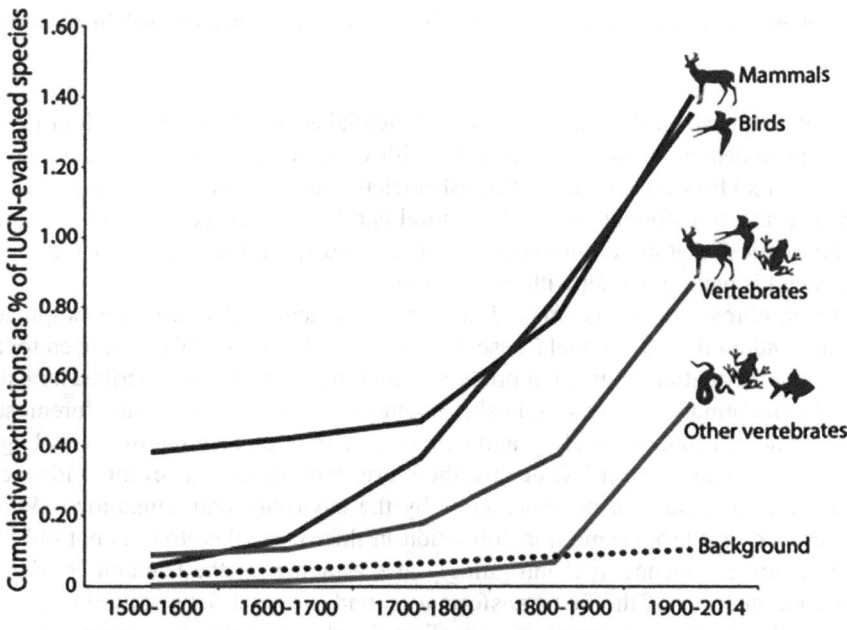

Figure 8.2 Vertebrate Species Extinction, in Ceballos et al.[22]

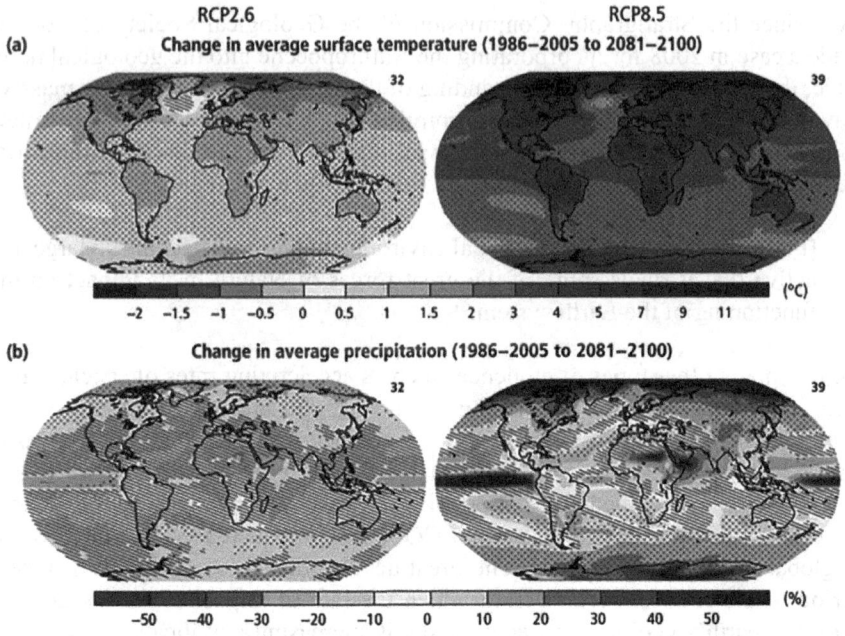

Figure 8.3 Change in Temperature and Precipitation, in IPCC Climate Change 2014, Synthesis Report, Fifth Assessment Report.

http://ar5-syr.ipcc.ch/ipcc/sites/default/files/AR5_SYR_Figure_SPM.7.png, accessed 16 November 2017[23]

Karl Polanyi had earlier analysed in his influential economic history back in 1944 an unprecedented change that occurred with what he in his book title described as "the great transformation" of English society into a technological market society. It was a transformation of "the natural and human substance of society into commodities" that are unlimited and that "must disjoint man's relationships and threaten his natural habitat with annihilation".[24]

Later, climate scientists adapted the term anew and declared, in a widespread memorandum from the Nobel Cause symposium in Potsdam 2007, "the need for a great transformation".[25] It is important to remember that what is described here as great transformation is not simply about natural processes "but first and foremost, a question of economy, society, and culture", as aptly pointed out by Claus Leggewie and Harald Welzer.[26] Crucially, therefore, extreme weather events and other challenges also need to be approached by the environmental humanities. What is meant by Anthropocenic transformation in this sense, therefore, is not only a subject for geoscience. It is intriguingly also a theme for the humanities where the consequences of the first transformation and the need for a second must be analysed as normative narratives as well as deeply impacting iconographies and global visual politics,[27] which, however, do not yet seem to have found their full

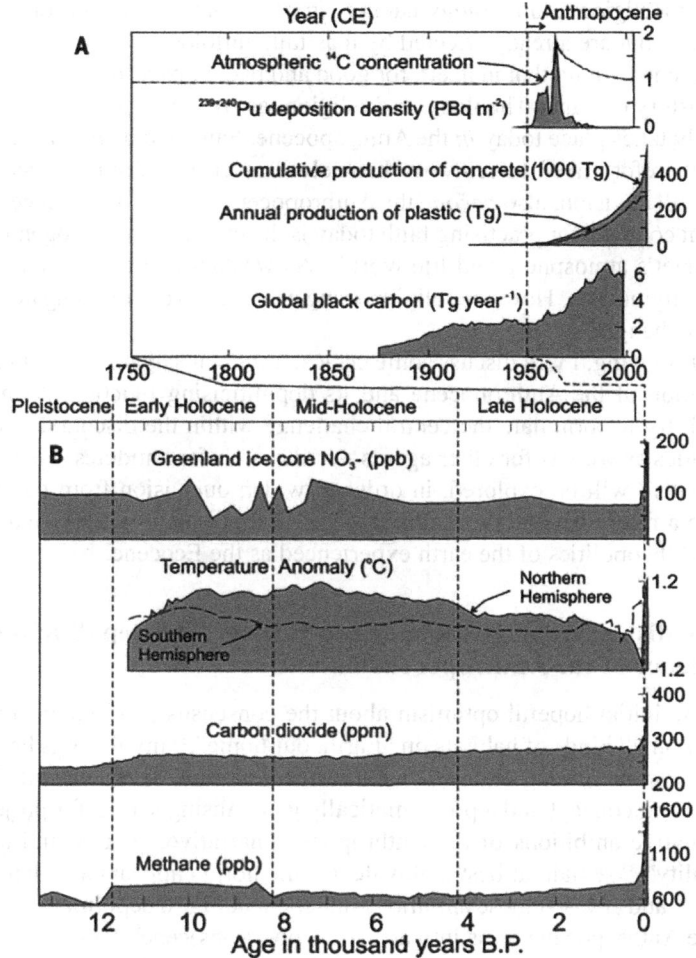

Figure 8.4 Key Markers of Anthropogenic Change, in Waters et al., figure 1.

strength. The Anthropocene also has an ethical dimension, as the whole of humanity is now to be held responsible for what happens on Earth, even if one naturally cannot ascribe this responsibility to all humans in general, as most of them have in fact not lived in unsustainable ways.[28]

While scientists have enthusiastically embraced the suggestion, which, it should be noted, was not developed by geologists but by self-critically aware geoengineers and chemists, others have criticised the implicit anthropocentrism and the practice of human eco-management as an almost God-and-nature-given imperative.[29]

How is religion in general, and Christian theology in particular, affected by this discourse? Religion and theology cannot simply relate *to* the Anthropocene and its discourse but are already affected by it as faith unfolds as a practical and ideological human activity that in itself, for good and bad, impacts on the environment and the history of Earth. Theology, and religion (as we have shown in our book[30]), necessarily takes place today *in* the Anthropocene. Faith and its rational reflective systems therefore need to reinvent themselves as critical driving forces *within* and, as I will envision, also *beyond* the Anthropocene. One of the most central and significant contexts for practising faith today is the ongoing anthropogenic change of the planet's atmosphere and life worlds. As weather and climate change, this changes religion also. How can religion bring a change to our ongoing negotiation about that change?[31]

In the following, I will discuss some critical arguments against the triumphalist interpretation of the Anthropocene and its depoliticising function. As a second step I will try to formulate the central challenge within the discourse – for faith communities as well as for other agents – and search for antidotes. Finally, theological skills[32] will be explored, in order to widen our vision from the past and present to a future beyond the Anthropocene. In this way, it will move towards a contextual theopolitics of the earth experienced as the Ecocene.

Triumphalist eco-management or eco-justice? The depoliticising ambivalence of the Anthropocene narrative

After some initial hopeful optimism about the consensus that humans today are impacting on all kinds of habitats on "Earth, our home",[33] my feelings have transmuted into an increasing ambivalence towards what now seems to function as a homogenising concept and a problematically generalising screen for projection.[34] The normative ambitions of the Anthropocene narrative, as a "grand narrative about reality",[35] remain at best ambivalent,[36] the notion appears as a Janus-faced character,[37] and at worst these ambitions rather encourage a depoliticising attitude, where the Anthropocene turns into an "Anthropo-(Obs)cene".[38] Or is the Anthropocene as geological and cultural narrative simply a great scientific saga about the future and potential futures?[39] One should not misunderstand my "simply" here, because the straightforward storytelling and the discourse itself already implies the manifestation of its own reality and thereby is able to create powerful practices, a performative process that we can observe clearly, for example, in social movements such as the global young people's *Fridays for Future*, where students in a very short time and in many places around the world are able to accumulate an enormous amount of power and influence over climate politics.[40]

Three critical points should be emphasised.

Firstly, will the insight into the all-embracing impact of humans lead to a new humility towards both human and other life forms, *or* will it fertilise a new triumphalist self-understanding of humankind and a utilitarian agenda with regard to human technocratic and economic management? Even if the introduction of the term, fortunately enough, has rather followed the humble path, one can trace among scientists a certain degree of a socio-engineering attitude to the different spheres of life. The narrative about the Anthropocene does not, to put it plainly, produce any antidotes against anthropocentric superiority and absolutism.

Rather, the opposite is the case, as the concept of the Anthropocene is deeply depoliticising, as Erik Swyngedouw has recently shown.[41] Its central postulates about humanity causing the great acceleration are misleading, as the majority of the planet's inhabitants have not partaken at all in the process of damaging the environment. What is depicted as "humanity" in the discourse instead concerns a small minority of nations that have enriched themselves at the expense of others. Nevertheless, recently the situation has become less black and white, and we can observe a changing stratification where the distinction between the environmental impacts of industrialised, BRIC, and developing nations has become more subtle.

While the environmental humanities, to which ecotheology and the studies of religion and the environment also belong, reflect on nature as a source of gifts and commons of life,[42] regarding the human as an integral part of nature, earth system analysis often, even if not in general,[43] operates with a poor reductionist understanding of the human and social, which stands in sharp contrast to its highly sophisticated model of complexity with regard to natural processes. While religions compress the narrative into the language of "respect towards", "wisdom about", and "compassion and wonder within" nature, science continues to take an external, somehow meta-physical position from which to describe nature. The fatal doctrine of nature as (a market (?) for) "ecosystem services", which is popular in economics and some earth sciences, builds furthermore on the illusion that all life exists mainly for the sake of humans. But is the sky really made for *us* (Fig. 8.5)?

One can also formulate a similar criticism with regard to climate scientists' presentation of their insights to the public, where the well-known hockey stick appears on almost all graphs to indicate the shift from a safe and comfortable to a dangerous and threatening state of being (Fig. 8.6).

Figure 8.5 Clouds over Kattegatt.

Photo: Author, October 2017

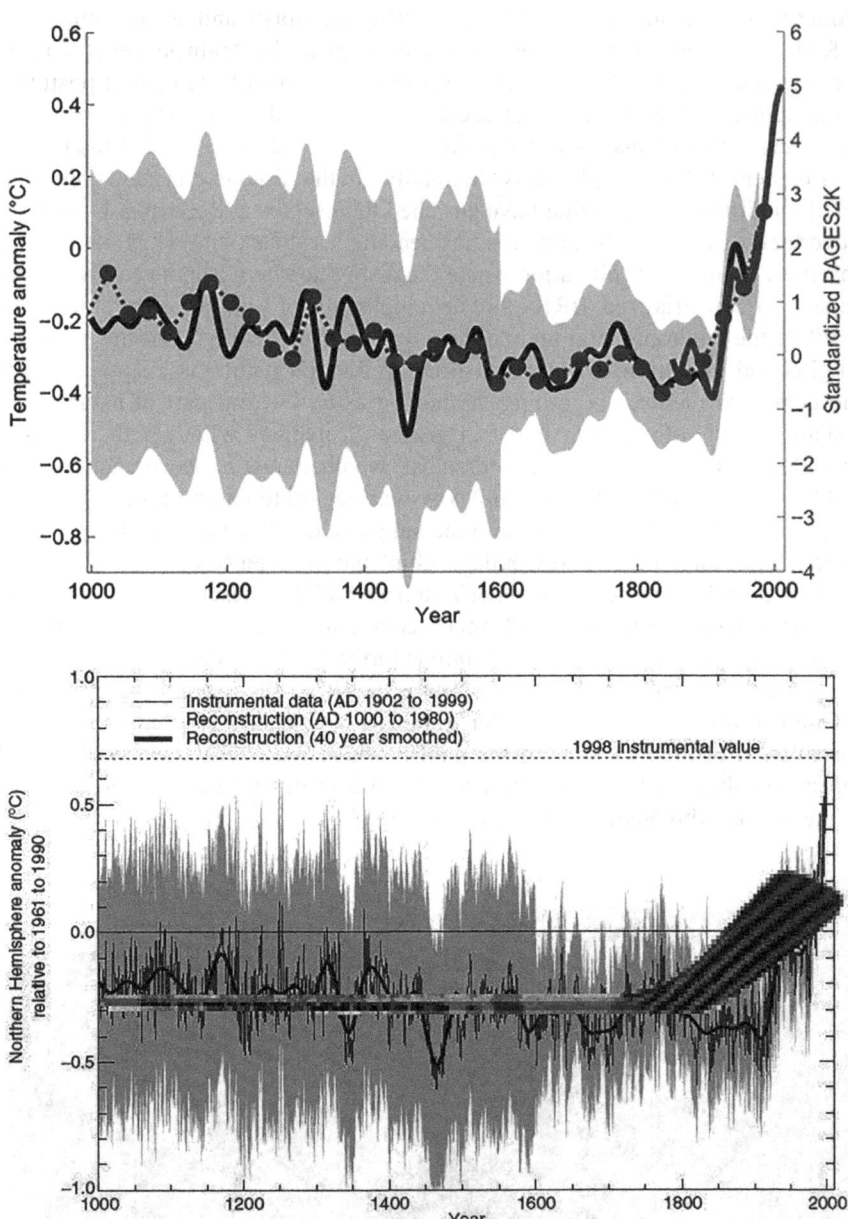

Figure 8.6 ab Green dots show the 30-year average of the new PAGES 2k reconstruction.
The red curve shows the global mean temperature, according to HadCRUT4
data from 1850 onwards. In blue is the original hockey stick of Mann, Bradley,
and Hughes (1999) with its uncertainty range (light blue). (Colors visible at the
website, see below.)

Images serve here, as Michael Mann has shown,[44] a normative function. They can at the same time function as mission statements, evidence, and guidelines, and they intend to convince the observer with their visual power in the same way as rhetoric. Images function as arguments, which cannot be separated from their intention to persuade. They act in order to present pure evidence. Considering the fact that they are, for the most part, based on computer-simulated models of real processes, one can rightly speak of an iconography that emerges from the scientist's and engineer's view of computers' naturalness. Nevertheless, the simulation, along with the computer, remains a man-made artefact, although its use can greatly assist in the empirical observation of nature.

Simulated worlds are represented, as Birgit Schneider explores in detail, as realities.[45] Is there at present, therefore, no alternative way of visually presenting the scientifically simulated climate futures? In any case, one should truly follow Schneider's and other media scholars' advice and take visual depictions of climate seriously but at the same time also constantly evaluate them critically. Such critical investigation differs radically, however, from climate scepticism, which enjoys exploiting such critical points deceptively (as for example the hockey stick) in order to misleadingly deconstruct and violate scientific knowledge. Religious aesthetics in this context without doubt has many alternative methods to offer for interconnecting realisation and simulation, as we will see later on.

A second criticism regards the lack of power analysis in the narrative. Andreas Malm and Alf Hornborg rightly accuse the narrative of neglecting the uneven distribution of wealth as a condition for the very existence of modern, fossil fuel technology, and of ignoring the fact that humans have caused global warming over the course of their long history.[46] Is Anthropocene thinking simply extending the natural scientists' worldviews to society?

Humans *in general*, in my view, do not exist; they always live and act in particular, in concrete contexts.[47] Given the fact that the majority of the planet's poor are suffering from the violence of ongoing climatic change caused by a minority of countries that have become rich at the expense of others, both human and non-human, one must ask if the Anthropocene narrative can include the necessary reflection about environmental and climate justice.[48] The naturalness of a consensus about the human impact in general tends to obscure the violation of justice, in the relational web of nature as well as in the asymmetry of world society.

The depoliticising function of the discourse makes it hard to simply embrace and accept it. Swyngedouw rightly characterises the postulate of a strange temporal rupture in modernity as "enrolling revisionist geo-history" that is silencing the controversies that have marked modernity. Conflicts that have promoted actual problems are ignored when past societies and their environmental awareness are simply denied. Rather than focusing on "a division of modernity between a before and an after", one needs to consider "a division *in* modernity".[49] The lack of power analysis and the depoliticising function of the discourse are further supported by different so-called radical new ontologies, which according to Swyngedouw have the function of "mak[ing] sure nothing really has to change". Talking about climatic change in the Anthropocene nurtures an illusion that we can adapt to change by establishing technological, political, and economic instruments that make sure that we – in our values, ideologies, lifestyles, and ordinary consumerist behaviours – do *not* (really) need to change. Is it enough to talk about "policy change" and will this also change the human deep down, or is Rilke's persuasive demand still valid: *Du mußt Dein Leben ändern* (You must change your life)[50] – which undoubtedly also leads to the unavoidable *Du mußt Dich selbst ändern* (You must change yourself)?

One can therefore aptly talk about a "de-politicized politics" of the "Anthropo-obScene" as "an immuno-biopolitical fantasy".[51] The vision of a safe and secure life without any risks is a dangerous fantasy, which Alain Brossat has circumscribed as "immunitary logic" that destroys the community and represents a politics that disavows the political.[52] Following Swyngedouw, therefore, one cannot find any code, injunction, or ontology in the Anthropocene discourse – or, for that matter, even in the other *-cene* narratives – that can found a new political ecology. Politics rather builds on interruptions, where the human divisiveness is acknowledged seriously, and where one strives for equality and freedom by interruptive acts. Envisioned in this way, the Anthropocene hinders rather than enhances and avoids rather than encourages interruptive political action. It not only lacks an analysis of power, as Hornborg and Malm have stated, but also actively disables it. We should therefore strongly demand a process of re-politicisation.

My third critical point focuses on the somewhat apocalyptic tone of the Anthropocene narrative. Have we really reached the end? Is the whole of the planet's future from modernity onwards at the mercy of the humans now and hereafter? Might there be a new *-cene* (Greek *cene* = recent, new) after the Anthropocene, and might there be other forces that impact on our common future and our common earth?

Questions like these seem impossible to ask within the current narrative. Humans are, as we have seen, generalised in a misleading way, and essential conflicts and injustices are obscured. Anthropocene talk might in its worst form serve as a fetishising instrument that precludes striving for change. Theologically, we must therefore state that it breaks with an essential understanding of eschatology. From a religious perspective, the future of the Creation must always remain open, for the Creator and for Creation's own power of evolution. Time, as well as space and place, cannot simply be confined by humans in the cage of technical models. However much computer monitoring is done in empirical scenarios, it is the bodily awareness and perception of our environment that remains significant for what we feed into the computers.[53]

Life as a gift cannot simply be turned into a scientific scenario and it can definitely not be turned into a commodity to be managed and exchanged along the well-known paths of fetishising capitalism. The method of fetishising and commodifying all that exists has certainly proven successful in the history of capitalist forms of exchange ever since antiquity, but the so-called price is destruction, injustice, violence, and a movement of the planet towards an "un-inhabitable" earth for most creatures.[54]

The narrative of the Anthropocene, and especially its depoliticising function, therefore leads to a radically new re-politicisation of the human condition of being alive. And it also demands a new understanding of visuality, as the production of images now moves into the centre of our understanding of the world and our-selves within it.[55]

From the perspective of the Abrahamic religions, with their belief in the world as a creation of the One, the challenge is deeply painful. How can one believe in a good Creator while his/her own creatures threaten Earth as a habitable place to live? "Is God angry"[56] at the humans, or does she/he still love them all and the poor most of all? How can one continue to believe in a good creation and envision a *creatio continua* and *creatio futura*, the coming world, in such a situation? How is the image of God connected to the (scientifically designed) images of the world?

Delineating the spiritual pain so sharply should sufficiently convince us not simply to regard the anthropogenic impact on the earth as a question of environmental ethics but to become aware that being alive in the Anthropocene implies a radically new challenge to reconstruct one's identity, worldview, and image of God. And it demands re-politicising interruptures that are able to make visible the divisions and injustice in interhuman and human–non-human relations. No more, no less.

Antidotes

My critical questions to the Anthropocene narrative should not be misunderstood. There is of course substantial value in a scientific consensus about the depth and scope of the human (or better, the prosperous part of humanity's) significance for the future course of the planet. Without such a general consensus, environmentalism can rarely develop its transformative strength. It is further important to acknowledge the variety and multivalence of the Anthropocene discourse.[57] My point nevertheless is that the present narrative's narrowness and the Anthropo(Obs) cene's problematically depoliticising function, which are supported by proclaimed new ontologies, which are not always as radical as they pretend to be,[58] must be overcome, and that faith communities and the environmental humanities seem to have a central role herein.

Among others I can imagine four antidotes to be used in order to bridge the abyss of the narrative's "lack of power, history and ethics".[59]

History and remembrance

Franz Mauelshagen and Dipesh Chakrabarty rightly state that the Anthropocene narrative affects our understanding of history in general.[60] Once more it reveals the clash between the cultures of science and the humanities in a reification of the past and the future.

Historically, the idea of an overarching impact of humanity on the earth is not new at all. Vladimir Vernadsky coined in 1938 the notion of "noosphere", inspired by Teilhard de Chardin. Before him, Catholic geologist Antonio Stoppani argued in 1873 for a new geological period. As early as 1784, Johann Gottfried Herder was aware of the human being as a climate-altering species. For him cultural history in general is the history of humans changing the climate around them, and – here he seems to be the first to have become conscious of the reciprocal impact – "the inhabitants themselves have changed with the climate". According to Herder, mankind represents

> a band of bold though diminutive giants, gradually descending from the mountains to subjugate the earth and climates with their feeble arms. How far they are capable of going in this respect futurity will show.[61]

Today we know all too painfully well how far humans are capable of going (and how incapable they are of stopping).

Alexander von Humboldt's understanding of nature as painting, *Naturgemälde*, and the earth's surface (nature's physiognomy) as the "face of the earth" offers a further early holistic view of the earth affected by the human imprint (Fig. 8.7).[62] While these early thinkers were able to integrate spiritual, human, and cultural dimensions into their fairly holistic approaches, contemporary earth scientists

Figure 8.7 Friedrich Georg Weitsch, *Humboldt and Aime Bonpland in the Valley of Tapia on the Foot of the Volcano Chimborazo*, 1810, oil on canvas, 100 × 71 cm, Schloss Charlottenburg, Berlin.

https://commons.wikimedia.org/wiki/File:Humboldt-Bonpland_Chimborazo.jpg, accessed 30 January 2020

instead operate with quite a poor understanding of the human.[63] Thanks to these early Enlightenment scholars, we can today perceive and understand the atmosphere not simply as nature but also as a cultural object.[64]

My first antidote therefore is to plead for a more complex historical understanding of the interaction of sociocultural and natural processes, as it has been developed recently in the new discipline of climate history. In particular, the discourse about memory and remembrance offers here a much deeper understanding of the dynamics and diversity of continuities and changes in history.[65] Climatic change offers in such a view a radically new *Erinnerungsraum* (space of remembrance).[66] The history of religions should also necessarily be included herein, as human ecology is intimately interconnected with religious practices, values, and worldviews. Images of the sacred impact directly on human thinking about and acting with and in nature. Secularist modern and late modern environmental practices are also driven by doctrinal forces, which again one can analyse by applying methods from cultural and religious studies.

Aesthetic wisdom

Another antidote regards the lack of self-critical thinking with regard to ethics. Much has been said and written about environmental ethics in the last 50 years, although unfortunately not much has been applied in environmental politics. Here I would only like to point to one contribution from epistemology, that is, the distinction between knowledge and wisdom.

Following Nicholas Maxwell,[67] science, as well as the humanities, is producing knowledge that is not simply the agglomeration of information into a computer model but a qualitative synthetic process of bringing together different kinds of observations and reflections. But how should one apply knowledge? And how should one select the preferable among the many insights? How could environmental knowledge turn into re-politicising practices that do not follow the Anthropo(Obs)cene narratives?

Following Maxwell, wisdom is the art of reflecting on how one should use knowledge. Wisdom includes rational reflections and moralities, and it also integrates worldviews and values that can be anchored in a religious and/or cultural context. With regard to nature, wisdom and wonder are deeply interconnected.[68]

Following Tim Ingold, it is the perception of our environment and the skills to become aware of what it means to be alive that are necessary to achieve meaningful negotiations about what we should or should not do. Liberation theology has in a similar way argued that seeing the poor represents a necessary presupposition for acting and thinking. The well-known circle *Seeing-Thinking-Acting . . . and Seeing again* has served as a central method for both pastoral and academic contextual theology, but it is in my view the third bridge, from *Acting* to *Seeing anew*, that has received far too little attention. Applied to the Anthropocene discourse, far more wisdom is therefore needed in evaluating the empirical insights of scientists, and a stronger aesthetic dimension is needed in perceiving and becoming aware of the suffering of the victims, both in nature and society, the "poor creatures" so to say.[69] In a spatially conscious eschatological key, we need to regard the victimised people, those who have no room to negotiate space, as the subjects experiencing the *Eschaton* (end time), the new world to come.[70] Those who are suffering in non-places and have no history are now no longer objects but subjects, and autonomous actors of change in times of Anthropocenic lack of freedom and inequality.

Complexity of the whole

A third antidote regards complexity and holism. Certainly, the environmental sciences, and especially climatology, which is a driving force within the Anthropocene narrative, interconnect a large variety of empirical data, methods, and theories. Earth system analysis represents in itself an enormous success of transdisciplinary research, and its results are impressively differentiated and clear.[71] Even if empirical observations are turned into artificial, human-made, computer simulations that are limited by engineering practices and thinking, one can without doubt rely on the conclusions.

Nevertheless, human beings, local and regional populations, sociohistorical developments, and unexpected so-called irrational behaviours are not included to a satisfactory degree in the same complex and differentiated way as the scientists' investigation of the atmosphere. Although climatology explicitly focuses on the anthropogenic impact, it seems to be fairly ignorant about the complexity of its cause, that is, the human, cultural, historical, and spiritual dimension of this impact.

In brief, the Anthropocene narrative deals with human impacts but unfortunately not much with humans, and it lacks as we have seen a crucial insight into the violent splits and controversies of and in modernity. The feedback impacts of climatic change on human responses are not included in the scenarios, nor is the complex diversity of human behaviour with regard to response to change adequately monitored. In the most recent IPCC reports, some ethical aspects have just started to be included, mainly due to the commitment of Catholic economist Ottmar Edenhofer and others.[72]

Furthermore, the concept of change in climate impact science seems philosophically to build on a rather limited and narrow understanding. Change is (as one can learn from Wassily Kandinsky's rhythmic composition (Fig. 8.8)) never simply a transfer from one state to another but a multifaceted, hard to predict, movement of variation (which one could approach, e.g., from becoming aware of the flux of weather with and beyond the modern meteorology that we have mined more deeply in the preceding chapters).

Social psychologist Harald Welzer has rightly recently criticised environmental organisations and the committed political green parties for a fragmentation of the complex interconnectedness of environmental processes. Environmentalist politicians often operate in what he calls a "reductive culture",[73] with a kind of problematic illusion of the technocratic doability of everything, and pretend that there are easy solutions. Let us reduce the number of cars; let us eat less meat,

Figure 8.8 Wassily Kandinsky, *Composition VII*, 1913, oil on canvas, 200 × 300 cm, State Tretyakov Gallery, Moscow.

http://uploads8.wikiart.org/images/wassily-kandinsky/composition-vii-1913.jpg, accessed 30 January 2020

increase solar energy, etc. Of course, we need not only an energy turn but also a mobility turn, a food production turn, an urban planning turn, etc. But as long as political agendas of the day focus only on selected issues, they will never achieve the broad and large-scale cultural revolution that is needed to avoid the earth (or better: even more places on Earth) turning into an uninhabitable place. For many regions, this turn is already in full swing, and climate and environmental conflicts and migration waves are accelerating. The call for repentance should therefore not be split up but remain unified and comprehensive.

Conversion

To use a classic word from the Christian tradition to summarise Welzer's conclusion, what we need is conversion. Biblical Greek *metanoia* aims at the conversion of the mind, but it certainly also includes the conversion of the eye, the body, and the social modes of existence. It does not offer an alluring way of thinking that does not require any real change but makes painful demands on us.

Especially in the Christian tradition, eschatology contributes to this ongoing social communication of imagined futures with a specific capacity to act: repentance and conversion. Both seem, if one follows the biblical sources, to have their place at the beginning of a believer's life in the eyes of God, as well as needing to be practised continuously in daily life, as everyday conversion. Calling upon Leonard Cohen's lyrics, where he asked[74] "When they said REPENT REPENT – I wonder what they meant", what do repentance and conversion mean?

Reading the gospels, the act of conversion is directly connected to a specific quality of time, in the present time of the early Christian believers, that is, the *kairos*, a time of challenge and conversion, connected to the cruelties and life-threatening social processes of the Empire. Matthew's

> [r]epent, for the kingdom of heaven has come near
>
> (Matthew 3:2, NIV)

is directly located in Mark's identification of the present time as *kairos*, a specifically challenging time of crisis as well as of future-shaping decision:

> The time (kairos) has come.
>
> (Mark 1:15, NIV)

The call to repentance is anchored in the Jewish faith, where it was central in the Hebrew Bible, and especially in the prophetic tradition. Repentance here was anything but an individual affair – it was a prophetic call for the whole people and a clear resistance to the fabrication of life- and faith-threatening idols.

A fourth antidote might be circumscribed as spiritual. Some theologians and representatives of faith traditions are arguing that religion in general offers salvation and solutions to problems, and that religious faith is necessary for transforming this world for the better. I am rather sceptical, as religions represent human practices, no more, no less, that can function constructively as well as destructively. Faith carries the same pros and cons as all other human practices, such as politics, culture, or economics, and can therefore be regarded neither as absolutely good nor for that matter as absolutely bad (as many secularist voices like to preach). Life-enhancing liberating and pathological life-threatening forces, rather, can walk hand in hand. In the tradition of Reformation theology, 500 years later one can certainly wonder what it means in the context of the environmental challenge to be both a sinner and justified: *simul iustus et peccator*. Can Christian sin-talk be of any value herein, and how should one apply it as a force for conversion and liberation?[75]

Nevertheless, faith is, due to its specific skills, highly fruitful for overcoming the Anthropocene narrative's danger of turning into just one more imperialistic and politically depoliticising grand narrative of the universe.[76] "Faith" offers in my view a better technical term than "religion" or "spirituality" as it does not pretend that religion is a clearly separated and isolated social subsystem, or that there is a separation of the spiritual from the bodily being-alive. Faith in the biblical sense implies vision, belief, hope, trust, and truthfulness, and it manifests as a skill to relate and communicate. It departs from the existence of the sacred, whether that be a God, spirits, or an inspired nature. What we regard as sacred impacts on our feelings, perception, thoughts, and acts.

Much could be said about the diverse contributions of faith communities and spiritual wisdom with regard to the environmental challenge, and here I will only refer to some recently published standard works, such as Ernst Conradie, Willis Jenkins, John Hart, Laura Hobgood and Whitney Bauman, and our Forum's volume on "Religion in the Anthropocene".[77] The new transdisciplinary research field of the environmental humanities is also producing a wide range of insights.

Different theological approaches to the debate about the Anthropocene are emerging at present. While some regard the Anthropocene as "a grammatised *con*text" with high relevance for doing public theology,[78] others are emphasising the ethical dimension. Eschatology offers, without doubt, certain tools for encountering the already discussed lacks of the narrative. While Marion Grau, for example, develops a "petro-eschatology" to overcome the uninhibited use of fossil energy, I have designed a spatial eschatology where the encounter with the life-giving Spirit in the topographies of social and environmental suffering moves into the spotlight.[79] Assisted by such a spatial eschatology, one can also overcome the narrow and apocalyptic timeline of the Anthropocene narrative, and envision the Ecocene beyond or possibly within the Anthropo(Obs)cene.

Another shared future – towards the Ecocene

My strongest objection to the Anthropocene narrative lies in the question of how we imagine what we might meet *beyond* the Anthropocene. Is there any space to imagine a new geological era beyond the Anthropocene? Or should we abandon the method of periodisation in general? Maybe the vision of an "Ecocene", where human and other life forms overcome their divisions and conflict and cohabit on Earth in fully just and peaceful entanglements, can be of help? Or rather an era of apocalypse where humans eradicate themselves from the planet, followed by an era of new genesis where evolution searches for new paths without human intervention? Or a "post-Technocene" where the fetishisation of money and machines has been overcome and technical spaces have turned into lived spaces?[80] How can one make sure that such a vision of the Ecocene does not simply turn into one more version of an Anthropo(Obs)cene, or into an Eco(Obs)cene?[81]

Is there any thought about the future in the narrative of the Anthropocene, or rather a total absence of utopia? While religions always operate more or less strongly with images of the future and so-called eschatologies, the narrative of the Anthropocene, as far as it is negotiated at present in the Anthropocene Working Group, seems to lack not only self-critical skills with regard to politics, power, history, and ethics but also the skill to imagine a future beyond the present. How could it thereby make politically evident its social and environmental relevance?

In my view, the notion of *Ecocene* might offer a more constructive term for a synthetic vision of an open, shared future. While terms like "capitalocene" or "post-Technocene" operate with a limited focus, sound too dystopian, and run the risk of obscenity instead of sincerity, the notion of Ecocene might enhance positive visions of the future and include entangled dimensions, envisioning

> an epoch in which not one single species determines the future of the Earth but the collective wisdom and interactions of all species and the bio- and geophysical world.[82]

The term is in the air, so to speak, and has most recently entered the discourse not so much in science but in other spheres such as architecture, design theory, and theology. It refers to a geological period beyond the Anthropocene, or rather a slow transformation from the one into the other, where the whole of the ecological sphere embraces and integrates the human. Design theorist Rachel Armstrong states that

> there is no advantage to us to bring the Anthropocene into the future. . . . The myth of the Anthropocene does not help us . . . we must re-imagine our world and enable the Ecocene.[83]

For biologist Robert Steiner it is

> inevitable that the current Anthropocene era will evolve into an ecologically sustainable era – which can be called the "Ecocene". The current trajectory of environmental and social decline cannot continue much longer.

Consequently, he focuses on the question of what comes beyond, as

> indeed, the Anthropocene will be gone in the blink of geologic time. The real question is: what will be left of the biosphere at the dawn of the Ecocene, e.g. what species, including Homo sapiens, will survive the Anthropocene evolutionary bottleneck?[84]

As we have already seen, the contemporary discussion about the Anthropocene suffers from a fatal lack of political and historical consciousness and a lack of including the future, or better different futures, in its narrative and iconography. While scientists at present mainly debate about the beginning of the so-called Great Transformation, and its main reasons and driving forces, their discussion of the future remains quite general. While some imagine it as an apocalypse and cosmic disaster, others regard it as a promising new arena for socio-engineering. The challenge in contrast is not to fall into either of these gaps, but to imagine and negotiate our *shared* future,[85] a just and sustainable future that can be shared equally and in freedom by all creatures.

It is probably this lack of a qualified reflection on potential sustainable and unsustainable futures that accelerates the triumphalist and depoliticising danger of celebrating this new period as a new period of human geo-management. This lack also seems to be part of a wider cultural shift where our ways of imaging the future (and the past) are undergoing a radical shift in the modern time regime.[86]

Such an interpretation immediately produces a deep conflict with faith, as the future in the Christian tradition must remain always open for the Creator and Liberator to act in. According to Jürgen Moltmann, God encounters his/her creation from the future, and I would add also from the past. Time and history, as well as space and place, always remain transparent for the Triune.

The Christian creed summarises this belief in its words about the new world (aeon) to come, and this *aeon* can scarcely refer to today's Anthropocene, a time of human mismanagement and an uninhabitable place for humans to live in as God's images. Rather, there is a deep need to hope for a new world to come beyond the Anthropocene.

While apocalyptic thinking imagines the end as a future of chaos and disaster, and manipulates and terrifies its audience, eschatology operates with an integrated present *and* future dimension. It is not simply the interconnection of the now and then, but also develops, as Vitor Westhelle and I have recently shown, as a spatial theory. The placial encounter with the God of the Here and There and the God of the Now and Then transforms the places in need of liberation. Climatic change represents such a place as it makes it necessary to encounter the Life-Giving Triune Spirit who takes place both now and then, and both here and there. Faith needs to be reconstructed, faith in the Spirit hovering over the vibrating waters of chaos in the beginning and the Spirit as the Giver of Life and as the source of the new creation to come.

Applying such a spatial and liberative eschatology to the narrative of the Anthropo(Obs)cene, it is impossible to imagine the future as a simple age of the humans. "Eschatology as imagining the end"[87] must necessarily stretch beyond the life-threatening anthropogenic impact that the rich nations have executed in the great transformation. Hope needs to flourish so that another age might appear.

In my view, such a vision of a re-politicised Ecocene fits perfectly with the biblical vision of a *creatio continua* that flows into a *creatio futura*, that is the (ongoing) creation of the new heavens and new earth. Biblical imaginaries for such an Ecocene are many: the heavenly Jerusalem as a truly eco-urban life-sphere, peaceful pastoral grazing of wild and other animals in the meadows, the pastoral vision of God as good shepherd and the people and creatures as a herd in a harmonious ecology, or the thanksgiving ritual after the flood and climate change disasters when the bow in the sky turns from a symbol of war to a colourful sign of peace between all created beings (Fig. 8.9).

Ethically, theologians have argued from such a vision for a future-oriented human ethics of responsibility, following Dietrich Bonhoeffer.[88] Personally I would prefer a discourse ethic model to a virtue or responsibility model.[89] In comparison with the three dominant ethic models (utilitarian, deontological, virtue ethic), elaborating discourse ethics with regard to demanding future moral problems avoids falling into the trap of stereotyped, one-sided monist constructions of the one normative guiding criterion, whether it be Utility and Consequence, Duty and Principle, or Virtue and individual moral Character. Rather, the communicative process of those who are impacted by the moral problem (which in our case are all human and non-human beings) needs to find a way of defining, negotiating, and solving the problem, gladly supported by background warrants in each of the three dominant models and other marginalised and non-Western models.[90] Global institutions such as the UN, UNEP, and IPCC already offer efficient social arenas and methods, which sadly are still misused by those superpowers

Figure 8.9 Joseph Anton Koch, *Das Opfer Noahs*, 1803, oil on canvas, 86 × 116, Frankfurt am Main, Städel Museum.

https://uploads7.wikiart.org/00139/images/joseph-anton-koch/jahrhundertausstellung-1906-katnr-0882.jpg, accessed 30 January 2020

and allies who place their self-interest over the planet's and humanity's common good (and its right to survive and evolve).[91] It thus appears as a priority today to entangle models and practices of human rights and environmental ethics, a theme outside the theme of this book, but certainly promising if one also starts to weather the field of human rights, which, for example, the section about Fiji in Chapter 3 touched upon. Human rights can only grow within the rights of nature.

In short, it is the vision of another period beyond the Anthropo(Obs)cene that allows a deepening and widening of the ambiguous Anthropocene narrative and a transformation into another period where humans are no longer masters and rulers of Creation but subjects of Creation, no longer shepherds but sheep in a flock, no longer alleged givers but true receivers of life.

To sum up, one of many substantial critical and constructive contributions to transforming the narrative of the Anthropocene is to nurture hope and to establish practices that manifest this hope for the Ecocene. It envisions Earth as a home where freedom, justice, synergy, and peace can flourish, a world such as in Giovanni Bellini's Renaissance paintings (Figures 8.10 and 8.11) that embraces most of the movements in God's history of salvation: the earth as a home where the Life-Giving Spirit is taking place.

Figure 8.10 Giovanni Bellini, *Sacred Allegory*, ca. 1485, oil on wood, 73 × 119 cm, Galleria degli Uffizi, Florence.

https://commons.wikimedia.org/wiki/File:Giovanni_Bellini_001.jpg, accessed 30 January 2020

Figure 8.11 Giovanni Bellini, *Saint Jerome Reading in the Wilderness* (detail), 1505, oil on canvas, 49 cm × 39 cm, National Gallery of Art, Washington.

https://en.m.wikipedia.org/wiki/St._Jerome_in_the_Desert_(Bellini,_Washington)#/media/File%3AGiovanni_Bellini_St_Jerome_Reading_in_the_Countryside.jpg, accessed 30 January 2020

Re-politicising the narrative and iconography of the Anthropo(Obs)cene in the frame of a still-to-be-designed Ecocene radically challenges environmentalism, and in the same sense theology, at its core. According to John B. Cobb Jr., theology is merely reflecting about what matters for Christians,[92] and alluding to Leonard Cohen, popular problems are what we all commonly struggle with: the problems of the people.[93] Contextual theology, therefore, is simply to reflect encountering the triune God in the midst of what matters among people's popular problems (sic!). Following Dietrich Bonhoeffer, God is not found where the church is but the church must move to the places where God acts; the church (and hopefully also all other religious faith communities) deserves its own right to exist and act only insofar as it is a community in its "being-for-others".[94]

Lived spaces of an anticipated Ecocene, where interruptive intervention and conversion might come true in changing weather, are such places. The challenge to faith communities and religious belief in general occurs in such a light, as does the challenge to environmentalism. It appears as concerned commitment to being alive in entanglements of freedom, justice, empathy, and respect in the flux of weather in lands where the light, wind, and weather are gifts to all creatures in communion with each other and where the Spirit as Giver of Life blows the winds of a time to come flourishing for the benefit of all.

Weather wonder – synergising affective and meteorological atmospheres

After our criticism of a depoliticising obscene Anthropocene narrative and imaginary and a plea for a re-politicising sincere Ecocene vision ahead of us, I will finally return in the light of this to a discussion about modern meteorology's central term "atmosphere". Might an expanded and deepened understanding of the atmospheric nurture and catalyse the vision of a shared future beyond a violated present?

The atmosphere that surrounds and supplies us must, as Bron Szerszynski rightly observes, be regarded as

> having its own kind of abiotic life, as a far-from equilibrium system with dissipative structures, a "concretised" whole with parts that resonate with each other.[95]

While earlier times and cultures had a rich imagination of the natural phenomena wherein weather arose, modernity pulls together these manifold images into a single entity. Where the sky, clouds, heaven, air, wind, breath, meteors, and blowing weather constituted the flux in our surroundings, the scientific notion of the atmosphere absorbs all these domains and " 'kills' the air, stripping it of animacy and meaning".[96] According to Szerszynski,

> weather was in effect turned into a laboratory artefact, was "brought indoors", in an attempt to tame its material and semiotic unruliness.[97]

While Szerszynski goes on to argue for a more complex understanding of life in the openness of air, a plea that also includes acknowledging classical Jewish faith where life is animated and the embodied soul is constituted by a process of breathing and movement, I will here contrast the scientific to the phenomenological understanding as it has been elaborated in environmental aesthetics. "Weather" is not only a natural phenomenon; "weather is a rhetoric".[98] Therefore, one must also be aware of how language creates weather and how talking about weather is deeply connected to the history of sincerity, as Lisa Robertson can show. For her, "the history of the description of weather parallels the history of sincerity as a rhetorical value".[99]

While weather in science, as we have seen, is defined as "the condition of the atmosphere at any particular time and place",[100] the atmosphere here is understood as a thermodynamic system that unfolds in constant change and variation that takes place on such a large scale that humans have to accept and respect its unpredictability. Environmental aesthetics, however, can allow us to take atmospheric thinking further.

> If the atmosphere is the medium in which life is lived, then the weather is our vernacular term for what goes on in it,

Tim Ingold strikingly states.[101] Atmosphere in aesthetics is understood in a different sense than in geoscience, and together with related notions such as "aura", "ambience", and "mood" (German *Stimmung*), it makes it possible to interconnect the outer with the inner of human life. Such an understanding of atmosphere can, as I showed earlier,[102] overcome the split between the subjective and the objective and allow us to become aware of all that is taking place in between the inner and the outer.

In his unique and pioneering work on *Mensch und Raum*, Otto Bollnow postulated in 1963 that 20th-century philosophy, with time at its centre, marginalises the problems of "the spatial affectivity of being a human" (räumliche Befindlichkeit) and of "the space that man/woman experiences and lives" (erlebter und gelebter Raum).[103] According to Gernot Böhme, aesthetics is a self-aware human reflection on one's living-in-particular-surroundings.[104] In his approach, human beings should no longer maintain a distance from nature but should rather seek to participate in it. An ecological aesthetics of nature is as much a subjective self-reflection on human identity as it is a reflection on that which surrounds us, and, in addition, on the distinction between humanity and its enveloping environment. The notion of *atmosphere* achieves in this view a central function that allows us to interpret the qualities of the environment and the human being's *sich-befinden*, that is, how one is emotionally and bodily situated.[105]

> Atmosphere means, in this context, a sensorial and affective quality widespread in space. It is the particular shade or tone that determines the way one feels his/her surroundings,

as Tonino Griffero further emphasises.[106]

The atmosphere emerges in the space between the outer surroundings of the human and his/her inner bodily-spiritual *Befinden*. An atmosphere is in no way a diffuse, unclear, non-determined, shallow, or subjective entity, but it offers us a notion that emphasises in an exciting way the interconnectedness of the inner and the outer, the bodily and the spiritual, the surrounding and the indwelling. The atmosphere appears as

> a vague ens or power, without perceivable and discrete boundaries, which we meet around us and which affects our lived body and even involves us.[107]

Experiencing atmospheres, in fact, dissolves the distinction between subject and object. It is the encounter and the interaction between them that becomes the focus for our meditation. Atmospheres emerge in natural surroundings as well as in built environments. Not only names but also building materials are able to create atmospheres; they can "evoke history, enhance the habitability of a city".[108] In cooperation with related notions such as "accord", "aura", "ambience", and "charisma", the concept of atmosphere challenges the understanding of nature, religion, and culture. Inspired by Hermann Schmitz, Böhme emphasises that

> [a]tmospheres are always spatially "without borders, disseminated and yet without place, that is, not localizable". They are affective powers of feeling, spatial bearers of moods.[109]

While geoscientists and meteorologists depart from a clear spatiality and objectivity of what they regard as atmosphere, philosophers operate with a much more differentiated concept of spatiality, which is regarded as surfaceless and frameless:

> Atmosphere in this sense is a frameless, indivisibly extended occupation of a surfaceless space.[110]

In his theory of "emotions as atmospheres", Hermann Schmitz understands emotions as spatially pouring out and bodily reaching atmospheres. In this way, the new phenomenology resists the mathematical depletion of the external world and the inclusion of all that is subjective into the inwardness of a "homo clausus".[111] For Schmitz, self-consciousness appears as self-ascription, in an emotional bodily being-affected in a specific surfaceless (*flächenlos*) spatiality.

Mónika Dánél strikingly summarises the richness of such an understanding of atmosphere where the aesthetic, social, and geocultural potentials are simultaneously present. On this understanding, atmosphere can be seen as aesthetic category, as multisensorial bodily experience, as *passibilité* (Mădălina Diaconu) calling forth reception, as energy with a powerful effect on the senses of olfaction and touch (Peter Zumthor), as the in-betweenness of emanation and perception that can be grabbed as an affective power.[112]

Ingold nevertheless points to the striking fact that the atmospheric in this aesthetical frame is characterised by a complete exclusion of weather. "Both meteorologists and aestheticians", Ingold observes, "from their respective sides, are inclined to claim that *their* particular sense of atmosphere is primary, and that the other's is merely metaphorical".[113]

Nevertheless, Hermann Schmitz is able to connect weather and emotions in his concept of atmosphere, where emotions appear as weather, that is, as surrounding, unpredictable, diffuse, confused, and fogged. Sorrow, depression, and sadness can feel the same as oppressive close weather. Emotional and weather atmospheres are in this sense closely interconnected.[114] They are both pouring out. Böhme also connects emotions as atmospheres to weather; atmosphere depicts for him not simply the air space but the emotional tinge of space. In his view, weather is a natural phenomenon with an autonomy that one can become aware of in a bodily subjective being-affected. We characterise weather through emotions. In a critical evaluation of Goethe, who ascribes to weather observation purely meteorological qualities, Böhme emphasises weather as a spatial modification that one has to understand as an emotional atmosphere. It represents a total impression (*Totaleindruck*).[115]

How can one bring both these concepts together and rethink them in a way that is both affective and meteorological? Drawing on the example of a balloon flight, Ingold formulates an argument that "we need to refill the atmosphere with the material stuff of air".[116] In such a sense, he aptly concludes, the world that we inhabit is "a world of becoming, of fluxes and flows or, in short, a weather-world".[117] In such a world, the living beings are not only temporal and spatial but also temperate.

> And temperament is just another word for *mood* – that is, for the way the atmosphere pervades every pore of a living being and lends affect to its actions.[118]

By breathing, therefore, we are staying alive within the atmosphere.

Bringing together the atmosphere of geoscience and aesthetics makes it necessary to interconnect the physical and the embodied spiritual. While science usually excludes, or rather operates with a reductionist view of, the human dimension, aesthetics, as we can see, keeps at a distance from the scientific perspective. Geoscience in this sense, on the one hand, needs to learn how weather, in the lens of atmospheric analysis, impacts on human bodies and souls as well as on the whole of human sociocultural being and history. The so-called affective atmosphere[119] offers, from this perspective, a distinct part of the planet's atmosphere. Environmental humanities and aesthetics, on the other hand, need to learn more about the givenness of an atmosphere in constant change.

Mădălina Diaconu convincingly reflects on the essential need for variation – one should not simply identify good weather with a sunny cloudless state of well-being. Without meteorological variation, living in constant good conditions can "end in monotony".[120] Weather, she observes, is far too complex and subjective for a traditional aesthetics of natural beauty,[121] an insight that atmospheric geoscience can certainly support with a large quantity of empirical data. Even if aesthetic philosophers have sometimes experimented with a kind of meteorological aerial aesthetics, nature, weather, and the atmosphere rather serve as screens for imagination (*Einbildungskraft*), such as in cloud watching, which is a mix of observing and projecting. Aptly, Diaconu states that

> to enjoy the weather often means to enjoy the view of wide landscapes and waterscapes, to take pleasure in panoramas from the top of a mountain, or to look down at remote earthscapes from a plane, but not to enjoy the weather *for itself.*[122]

Such a view accordingly resonates with the fact that it may not be climate change itself that will be mostly perceived by the populace, but rather the impacts that it produces on our daily lives (in terms of reductions in fresh water, crop harvest losses, flooding, heavy rain, heat waves, extended periods of wet, cold, and cloudy weather, etc.). Diaconu furthermore notes how the ideal of fine weather has obviously evolved over a long cultural history; in late antiquity Church Fathers, such as Saint Ephrem the Syrian, were already describing paradise in a fictitious meteorological vision where the atmosphere is temperate, shaping ideal conditions for the fertility of the soil and "where the months' tempests are overcome" so that they cannot "pollute the glorious air".[123]

Reflecting on the atmosphere in aesthetic contexts, whether it be landscape painting, literature, or everyday clichés about fine weather, is closely connected to aesthetic modes of experiencing the quality of life. Weather, without doubt, has a large part herein, and the alteration of weather, which is at the core of meteorology, undoubtedly has an important lesson to teach about quality of life on a planetary scale as well as on subjective individual bodily and spiritual scales. Remembering our introductory reflection on the extreme heat weave and dry summer of 2018, the skill of aesthetic judgement becomes weakened: diffuse, vague, and blurred. Is "fine weather" really warm and hot? Does the absence of rain and a blue sky express beauty? Is "bad weather" always bad (for me and other creatures)? Might a deeper emphasis on reading alteration meteorologically also catalyse a richer aesthetic awareness of the mystery of weather? Meteorological and affective atmospheric thinking do not necessarily oppose but can also complete each other. Artisans and environmental arts might play a crucial role here in mediating visualisations of earth science in a way that enlightens science itself and at the same time inspires truthful imaginations of what is going on in ordinary life worlds.[124]

In another subtle and demystifying essay, Diaconu detects the embarrassing lack of communication about the atmosphere between scholars in meteorology and philosophy, which is also at the core of this chapter. This becomes even more fatal in the context of our contemporary post-postmodern longing for moods (*Sehnsucht nach Stimmungen*) where our so-called event society cultivates its passion for pure appearance. The notion of the atmosphere therefore refers not only to a scientific and an aesthetic but also to a sociocommunicative phenomenon.[125]

Up to this point, there is no sign of any convincing bridging of the gaps between these three modes of atmospheric thinking. While scientifically precise descriptions of the atmosphere, on the one hand, are not present in approaches of aesthetics of nature, on the other hand, meteorologists seem to have entirely missed the emotional, spatial qualities in aesthetic phenomenological atmospheres. "The whole pallet of moods [Stimmungen] that weather conditions are evoking within us and upon which artistic designs can be based" are disregarded in meteorology.[126] And if the subjective dimension announces itself, it is often reduced to selected appearances lauded for their classical beauty.[127] In her concluding discussion of open questions, Diaconu makes us aware of the challenge to intertwine the aesthetically descriptive with the aesthetically normative dimension of atmospheric thinking, and, in addition to this, also of the difficulty of criticising this without damaging the experience of an atmosphere. With regard to the event society also, atmosphere theory still remains ambivalent, as it can either fruitfully question and defog the economy's aestheticisation or catalyse it even more. "The implicit ethics of such a theory remains open", Diaconu states aptly, and points forward to what is, in my view, a crucially significant path, including for my own work, that is, to develop aesthetic atmospheric thinking anchored in the practised sensibility and quality of the encounter with the other, or *the strange/r* as I unfolded it earlier.[128] Compassion would then be able to serve as a central quality for atmospheric feeling, perceiving, thinking, and acting.[129]

For meteorological scientists, artists, event-addicted and ordinary open-minded/sensed people, the synergy of the perceived and the analysed, the subjective and the objective, of skies, places, felt and measured temperatures, etc. would then offer an entangled natural-and-sociocultural arena for encounters.[130] The atmosphere itself, so to say, might then offer us a lived and inspirited space for exchanging different modes and methods for exploring how we are carried (Getragensein) in our being-alive in the flux of weather. If it is permitted to confuse the reader even more, might the atmosphere, in such a physical-aesthetic-and-sociocommunicative sense, work as a kind of *Gesamtkunstwerk* (total work of art), that is, the planet's art work of weathered life-giving conditions interacting in synergy with the earthlings' sociocultural, spiritual, and technical artefacts?[131]

One might wonder if and how my argument might fit into the dynamically flourishing discourse about the so-called *new materialism(s)*, but due to my indifference to and my still-critical distance from some voices herein and what I consider the all too wide-open elastic range of the discussion, I must leave the reader in uncertainty. Nevertheless, I would like to point out a direction in which the synthetic atmospheric thinking suggested here might make sense. If new materialist approaches are successfully shifting the focus towards matter and material bodies in their dynamic becoming, and if they strive for a method of "reading insights through one another",[132] the meteorologically regarded atmosphere of course represents some kind of megamaterial and all-embracing structure that can be "read" by us at the same time as it can also "read" us. Weather sees/feels/reports us while we at the same time are seeing/feeling/reporting weather. And if interlocutors in the discourse really envision a communicative process far beyond the gap between natural and sociocultural sciences (a vision that I share but nonetheless cannot yet verify in my colleagues' science and technology faculties at present), the plea for synthetic and multifaceted atmospheric thinking about the Anthropocene on its way to the Ecocene would also make a lot of sense.

If, furthermore, new materialists develop their vision critically and constructively, as does Catherine Keller, who, inspired by the theory of quantum physics, has coined the notion of "entangled difference" in order to resist monist forms of resistance and to enhance an irreducible pluralism that is necessary for a fruitful strengthening of social interaction in a "bottomless ethics",[133] Diaconu's and my vision of an empathic, compassionate atmospheric feeling and acting for and with the weathered other/strange might also receive more strength. And if, finally, theological, spiritual, and ethical arguments such as Whitney Bauman's can be given, developing a new model of power-with rather than power-over,[134] that is, power that "lies with multiple-earth others in the evolving planetary community" such that it "means that whatever world we hope to bring about can only become through working with these earth others",[135] then in our striving for Ecocenic weather wisdom, in the sense presented in this book, we can also learn from weather-reporting-us how to feel, perceive, think, and act with and through each other in a true process of *inter-carnation*.[136] The flux of weather would then make and keep our body/souls/communities alive from above, below, within, and throughout. Wind and wisdom will blow and inspirit life into Earth and/in/at/for the earth's inhabitants on atmosphere-nurtured Earth, our home.

Especially in the frame of the discourse about the Anthropocene, such increasing production of meteorological knowledge in synergy with aesthetic awareness of the atmosphere in the flux of changing weather lands appears to me as necessary in the process of re-politicising the Anthropocene discourse in its turn to the Ecocene. If meteorology can inform the human, the political animal, about changing patterns of alteration, and if aesthetically aware human beings explore this alteration in the betweenness of subjects and objects, one might hope for a new political and ecological aesthetics of the environment. Even if scientific meteorology over the course of its history has gradually decreased the significance of the human observer,[137] one can hope for a coming increase in the significance of human observation of the meteorological instruments and results.

Empathy and weather-wising would then not just turn into individualistic esoteric practices but gain a strong transformative function. Sharing experiences about what changing weather does to me, to us, and to our common Earth would then produce new conditions for interruptive political action. Faith might in such a frame become a life-enhancing activity in synergy with the Creator Spirit in a frameless and surfaceless space. The science of the atmosphere could in such a space embrace insights in atmospheric aesthetics. Both together might found an awareness about longing and belonging, where humans are perceived as belonging inseparably to the earth's atmosphere as it constantly changes, and where humans as weather impacters begin, with the assistance of both meteorology and aesthetics, to become aware of how atmospheric impacts are impacts on themselves. Atmospheric belonging to the atmosphere would then turn into longing for another atmospheric way of being. Wisdom about what to do with such insights from the inside and outside presupposes the skill of wondering. These skills certainly should not lead to simple blue-sky adoration or "sublimity of catastrophes" consumed in a disaster tourism that turns weather into a commodified media spectacle.[138] Weather wisdom, then, as it has been explored in the frame of this book, necessarily demands new critical models and root metaphors that allow us to move forward on multiple scales towards Ecocenic life. The competence and "creative work of up- and down-scaling"[139] achieves new significance here, where architects (like those discussed in Chapter 7) and climatologists can meet and together educate and train the populace in a new global space. In the lens of faith, such a process of creative adaptation driven by empathy and creative wisdom presupposes belief in a God of surprise and the ability to wonder.[140]

Being alive in weather lands provokes such wonder every hour, every day, every week, forever. Bothering Goethe one last time by once more replacing "nature" in his quote with "weather",

> Do we only know ourselves as far as we know the weather, which we only become aware of in ourselves, and do we only in the weather become aware of ourselves?[141]

If it is only in the flux of weather that we can become aware of ourselves, a spiritually humble attitude to weather wonder and weather wisdom, fatally ignored too long,[142] will need to become a crucial skill for creative adaptation in a shared future on Earth. Wonder, hope beyond despair and weather world wisdom, *Wetter- und Weltweisheit*, would then turn into a matter of survival for us all in an increasingly volatile world and climate.

Notes

1 Jane Austen, "Letter to Cassandra," in: *Jane Austen's Letters to Her Sister Cassandra and Others*, collected and edited by R.W. Chapman, 2nd edition, London, New York and Toronto: Oxford University Press 1952, Digital Library of India Item 2015.533188, https://archive.org/details/in.ernet.dli.2015.533188/page/n9, accessed 2 December 2019.

2 Lesley Adkins and Roy Adkins, *Eavesdropping on Jane Austen's England: How Our Ancestors Lived Two Centuries Ago*, London: Hachette Digital 2013.

3 Friedrich-Wilhelm Gerstengarbe and Harald Welzer, *Zwei Grad mehr in Deutschland Wie der Klimawandel unseren Alltag verändern wird*, Frankfurt M.: Fischer 2012.

4 www.wma.net/what-we-do/public-health/climate-change/. See especially the WMA Declaration of Delhi on Health and Climate Change adopted by the 60th WMA General Assembly, New Delhi, India, October 2009: www.wma.net/wp-content/uploads/2016/11/HB-E-2015-1.pdf, accessed 2 December 2019. Cf. also the growing interest in the impact of climate change on mental health, for example in Pihkala's work on "climate anxiety". Panu Pihkala, *Climate Anxiety*, Helsinki: MIELI Mental Health Finland, Summer 2019, and Panu Pihkala, "Theology of 'Eco-Anxiety' as Liberating Contextual Theology," in: Sigurd Bergmann and Mika Vähäkangas (eds.),

Contextual Theology: Skills and Practices of Liberating Faith, London and New York: Routledge 2020.

5 The official definition of a heat wave according to the WMO is "more than five days in a series with a maximum daily temperature of more than five degrees over what is normal for the season in a period from 1961 to 1990". The Swedish SMHI defines a heat wave as "a period with a maximum temperature of more than 25 degrees on five days in a series".

6 And as if this was not enough, in the year 2019 – when the world's attention zoned in on the tropical rain forest fires caused by insatiable greed for profit in the alliance of Brazilian president Bolsanaro and capital-strong landowners, which threatened the climate, environment, and cultural justice alike – Arctic fires started in June and continued for weeks on a scale of what the WMO described as historically "unprecedented" events, where more than 100 wild fires ravaged the area and emitted an estimated 100 megatons of carbon dioxide into the atmosphere between 1 June and 21 July.
   Cf. https://public.wmo.int/en/media/news/unprecedented-wildfires-arctic and https://edition.cnn.com/2019/07/24/world/wildfires-arctic-climate-sci-intl/index.html, accessed 27 July 2019.

7 Roni Horn, *Weather Reports You: A Project of VATNASAFN/Library of Water*, Stykkishólmur, Iceland, London and Göttingen: Artangel/Steidl 2007; cf. www.artangel.org.uk/project/library-of-water/ and https://grapevine.is/mag/articles/2014/09/17/funding-dries-up-at-the-library-of-water/, accessed 27 November 2019.
   Departing from the idea that "everyone has a story about weather", which "may be one of the only things each of us holds in common" (p. 9), the project collected throughout 2005 and 2006 an archive of weather reports and stories from Icelanders in several places, beginning with the first place where in 1845 weather data were measured and regularly recorded in Iceland for the first time: Stykkishólmur. *At the same time we are reporting weather, weather reports us* turns into a pearl of wisdom that will continue to grow within me from this late November afternoon in 2019, in a plaguesome twilight, at what is hopefully the end of a long period of dismal cloudy, grey, wet, windy, and, in comparison with long-term measurements, unusually and uncannily warm weather, under an apparently boundless layer of clouds that seem as heavy as stones, in cheerlessly constant inflowing low-pressure fronts from the east, south, and west that pressure and burden body, soul, and spirit before light is again predicted to return tonight with northerly winds, negative temperatures, and the year's first snow, to the north of Lund in Southern Sweden, surrounded by what Rilke after his visits (at more comfortable times of the year) accurately described as *von Meeresahnung umgeben* (surrounded by [Baltic] sea sensations). Rainer Maria Rilke, *Briefe an Ernst Norlind*, ed. by Paul Åström, Partille: Paul Åströms Förlag 1986, 29 (Viareggio bei Pisa am 14. Juni 1904 an Ernst Norlind in Borgeby). While Horn turns weather into a subject, reporting on us, Braungart accordingly refers to the poet turning nature into the author-subject writing a poem for us. Georg Braungart, "'Die Natur, ich bitte um Nachsicht, schreibt ein Gedicht': Das Wetter in der deutschsprachigen Gegenwartslyrik," in: Georg Braungart and Urs Büttner (eds.), *Wind und Wetter: Kultur – Wissen – Ästhetik*, Paderborn: Wilhelm Fink/Brill 2018, 279–303.

8 Quoted in Stefan Rahmstorf, *Hitze ohne Ende*, Blog: Klimalounge, 6 August 2018, https://scilogs.spektrum.de/klimalounge/hitze-ohne-ende/: "Wir erleben diesen Sommer an vielen verschiedenen Orten der Welt zeitgleich ein gehäuftes Auftreten von Hitzewellen und extremen Starkniederschlägen. Genau eine solche Häufung von meteorologischen Extremereignissen wurde von uns als Folge des anthropogenen Klimawandels prognostiziert".

9 Michael Mann quoted in Damian Carrington, "Extreme Global Weather Is 'the Face of Climate Change' Says Leading Scientist," *The Guardian*, 27 July 2018.

10 Dim Coumou, "Rossby Waves and Surface Weather Extremes," *RealClimate: Climate Science from Climate Scientists*, 10 July 2014, www.realclimate.org/index.php/

archives/2014/07/rossby-waves-and-surface-weather-extremes/, accessed 12 October 2018.

11 J.A. Screen and I. Simmonds, "Amplified Mid-Latitude Planetary Waves Favour Particular Regional Weather Extremes," *Nature Climate Change* 4, 2014, 704–709, doi: 10.1038/NCLIMATE2271.

12 The middle latitudes of Earth lie between 23°26′22″ and 66°33′39″ north, and between 23°26′22″ and 66°33′39″ south, between the tropics and the polar circles.

13 D. Coumou (1, 2), G. Di Capua (1, 2), S. Vavrus (3), L. Wang (4) and S. Wang (5), "The Influence of Arctic Amplification on Mid-Latitude Summer Circulation," *Nature Communications*, doi: 10.1038/s41467-018-05256-8, published online 20 August 2018.

14 On the Rossby waves in the jet stream, cf. Rahmstorf, op. cit.

15 V. Petoukhov, S. Rahmstorf, S. Petri, and H.J. Schellnhuber, "Quasiresonant Amplification of Planetary Waves and Recent Northern Hemisphere Weather Extremes," *Proceedings of the National Academy of Sciences* 110, 2013, 5336–5341, doi: 10.1073/pnas.1222000110.

16 John Abraham, referring to a momentous climate study on the troposphere in "Scientists Detect a Human Fingerprint in the Atmosphere's Seasonal Cycles," *The Guardian*, 23 July 2018.

17 Press release from the Potsdam Institute for Climate Impact Research, August 6, 2018: www.pik-potsdam.de/news/press-releases/planet-at-risk-of-heading-towards-irreversible-201chothouse-earth201d-state.

18 Will Steffen, Johan Rockström, Katherine Richardson, Timothy M. Lenton, Carl Folke, Diana Liverman, Colin P. Summerhayes, Anthony D. Barnosky, Sarah E. Cornell, Michel Crucifix, Jonathan F. Donges, Ingo Fetzer, Steven J. Lade, Marten Scheffer, Ricarda Winkelmann, and Hans Joachim Schellnhuber, "Trajectories of the Earth System in the Anthropocene," *PNAS*, published ahead of print 6 August 2018, doi: 10.1073/pnas.1810141115, edited by William C. Clark, Harvard University, Cambridge, MA, and approved 6 July 2018.

   According to the World Meteorological Organization (WMO), the year 2019 was the second hottest year on record. For details, see https://public.wmo.int/en/media/press-release/wmo-confirms-2019-second-hottest-year-record (published 15 January 2020).

19 Steffen et al., "Trajectories of the Earth System in the Anthropocene," 7.

20 For the present state of the debate and the understanding of "the Anthropocene," see Colin N. Waters, Jan Zalasiewicz, Colin Summerhayes, Anthony D. Barnosky, Clément Poirier, Agnieszka Gałuszka, Alejandro Cearreta, Matt Edgeworth, Erle C. Ellis, Michael Ellis, Catherine Jeandel, Reinhold Leinfelder, J.R. McNeill, Daniel deB. Richter, Will Steffen, James Syvitski, Davor Vidas, Michael Wagreich, Mark Williams, An Zhisheng, Jacques Grinevald, Eric Odada, Naomi Oreskes, and Alexander P. Wolfe, "The Anthropocene is Functionally and Stratigraphically Distinct from the Holocene," *Science* 351, 2016, aad2622, doi: 10.1126/science.aad2622.

21 Will Steffen, Jacques Grinevald, Paul Crutzen, and John McNeill, "The Anthropocene: Conceptual and Historical Perspectives," *Philosophical Transactions of the Royal Society A*, 369, 2011, 842–867, doi: 10.1098/rsta.2010.0327, 842.

22 Gerardo Ceballos, Paul R. Ehrlich, Anthony D. Barnosky, Andrés García, Robert M. Pringle, and Todd M. Palmer, "Accelerated Modern Human-Induced Species Losses: Entering the Sixth Mass Extinction," *Science Advances* 1, 5, 19 June 2015, e1400253, doi: 10.1126/sciadv.1400253.

23 Cf. IPCC's special report from 2019 about what a global warming of 1.5°C implies: www.ipcc.ch/sr15/, accessed 27 November 2019.

24 Karl Polanyi, *The Great Transformation*, Boston, MA: Beacon 2001 (first published 1944), 44.

25 H.J. Schellnhuber (ed.), "Global Sustainability: A Nobel Cause," Nobel Cause Symposium, Potsdam Memorandum: The Need for a Great Transformation. Main Conclusions

from the Symposiums, Potsdam, Germany, 8–10 October 2007 [Potsdam Institute for Climate Impact Research (PIK), Potsdam, 2007].

26 Claus Leggewie and Harald Welzer, "Another 'Great Transformation'? Social and Cultural Consequences of Climate Change," *Journal of Renewable and Sustainable Energy* 2, 2010, doi: 10.1063/1.3384314, https://aip.scitation.org/doi/pdf/10.1063/1.3384314?class=pdf.

27 Cf. Birgit Schneider, *Klimabilder: Eine Genealogie globaler Bildpolitiken von Klima und Klimawandel*, Berlin: Matthes & Seitz 2018. On the Anthropocene as narrative, see Hoiß's critical discourse analysis: Christian Hoiß, "Das Anthropozän: Auf den Spuren einer Narration," in: Sabine Anselm and Christian Hoiß (eds.), *Crossmediales Erzählen vom Anthropozän: Literarische Spuren in einem neuen Zeitalter*, München: Oekom 2017, 13–37.

28 Cf. Konrad Ott, "Verantwortung im Anthropozän und Konzepte von Nachhaltigkeit," in: Rosa Sierra and Anahita Grisoni (eds.), *Nachhaltigkeit und Transition: Konzepte/Transition Écologique Et Durabilité: Concepts*, Frankfurt M. and New York: Campus 2018, 141–188.

29 Cf. Bonneuil's description of four different Anthropocene narratives at work at present and Pearson's emphasis on moving the discourse from the age's beginning to the "possibility of endings" (9), from the apocalyptic to the eschatological. Christophe Bonneuil, "The Geological Turn: Narratives of the Anthropocene," in: Clive Hamilton, François Gemenne, and Christophe Bonneuil (eds.), *The Anthropocene and the Global Environmental Crisis: Rethinking Modernity in a New Epoch*, Abingdon, Oxon: Routledge 2015, 18–23.
   Clive Pearson, "Is It Too Late? Doing Theology in the Anthropocene," 2015, www.otago.ac.nz/ctpi/otago419801.pdf, accessed 19 November 2017.

30 Celia E. Deane-Drummond, Sigurd Bergmann, and Markus Vogt (eds.), *Religion in the Anthropocene*, Eugene, OR: Wipf & Stock/Cascade 2017.

31 Cf. Sigurd Bergmann, "Climate Change Changes Religion: Space, Spirit, Ritual, Technology – through a Theological Lens," *Studia Theologica – Nordic Journal of Theology* 63, 2, 2009, 98–118, doi: 10.1080/00393380903345057.

32 For the understanding of theology as "skill", see Sigurd Bergmann, "Religion, Culture and God's Here and Now: *Contextual Theology* in Dialogue with Social Anthropology," *Svensk Teologisk Kvartalskrift* 81, 2, 2005, 67–76, https://journals.lub.lu.se/STK/article/view/6623/6243.

33 Earth Charter Commission 2000, *The Earth Charter*, www.earthcharterinaction.org, accessed 25 May 2015.

34 For a more detailed discussion of the criticism, see Sigurd Bergmann, "Is There a Future in the Age of Humans? A Critical Eye on the Narrative of the Anthropocene," Consortium for the Study of Religion, Ethics, and Society Forum, Spring 2016, Indiana University, www.iu.edu/~csres/csres_archive_14Dec2016/www/forum.php#Bergmann, accessed 19 November 2017.

35 Hoiß, op. cit., 19.

36 Celia E. Deane-Drummond, Sigurd Bergmann, and Markus Vogt, "The Future of Religion in the Anthropocene Era," in: Deane-Drummond, Bergmann, and Vogt (eds.), op. cit., 1–15, 14.

37 Sverker Sörlin, *Antropocen: En essä om människans tidsålder*, Stockholm: Weyler 2017, 209.

38 Erik Swyngedouw, "Interrupting the Anthropo(Obs)cene," opening key note, Munich, Deutsches Museum 17 October 2018, at the conference "(Um)Weltschmerz: An Exercise in Humility and Melancholia," arranged by ENHANCE Marie Skłodowska-Curie Innovative Training Network, co-sponsored by the Rachel Carson Center and the Deutsches Museum.

39 Cf. Helmuth Trischler, "Das Anthropozän: Neue Narrative zu Vergangenheit, Gegenwart und Zukunft," *Aviso: Zeitschrift für Wissenschaft und Kunst in Bayern* 3, 2016, 10–13.

40 www.fridaysforfuture.org/, accessed 28 November 2019.

41 See Swyngedouw, op. cit.

42 For a detailed theological discussion of the "commons", see Catherine Keller, Elias Ortega-Aponte, and Melanie Johnson-DeBaufre (eds.), *Common Goods: Economy, Ecology, and Political Theology*, Oxford: Oxford University Press 2015.

43 An increasing number of Earth scientists, however, seem to be become aware of this, and some are engaging in more holistic approaches; see, for example, the LOOPS activity: www.pik-potsdam.de/research/projects/activities/copan/loops, and its Special Issue: www.earth-syst-dynam.net/special_issue18.html, accessed 27 November 2019. For a reflection on "deeper human dimensions," including religious ones, see Dieter Gerten et al.'s thought-provoking and forward-looking argument: Dieter Gerten, Martin Schönfeld, and Bernhard Schauberger, "On Deeper Human Dimensions in Earth System Analysis and Modelling," *Earth System Dynamics* 9, 2018, 849–863, doi: 10.5194/esd-9-849-2018.

44 M.E. Mann, R.S. Bradley, and M.K. Hughes, "Northern Hemisphere Temperatures during the Past Millennium: Inferences, Uncertainties, and Limitations," *Geophysical Research Letters* 26, 1999, 759–762.
    Cf. also Michael E. Mann, *The Hockey Stick and the Climate War*, New York: Columbia University Press 2012.

45 Schneider, op. cit., 234.

46 Andreas Malm and Alf Hornborg, "The Geology of Mankind? A Critique of the Anthropocene Narrative," *The Anthropocene Review* 1, 1, 2014, 62–69, 64.

47 Cf. Sigurd Bergmann and Mika Vähäkangas (eds.), *Contextual Theology: Skills and Practices of Liberating Faith*, London and New York: Routledge 2020.

48 For a theological exploration of social and environmental justice, see Rasmussen's elaboration of "creation justice". Larry L. Rasmussen, "From Social Justice to Creation Justice in the Anthropocene," in: John Hart (ed.), *The Wiley Blackwell Companion to Religion and Ecology*, Hoboken, NJ: John Wiley & Sons 2017, 239–255.

49 Neyrat 2016, 117, cited and translated by Swyngedouw, op. cit.

50 Rainer Maria Rilke, in his poem *Archaïscher Torso Apollos* from 1908. Cf. Peter Sloterdijk's essay from 2009 with the same title *Du mußt dein Leben ändern* but subtitled *Anthropotechnik*, wherein he fatally transfers Rilke's tremendously life-giving struggle to deepen the human's interaction and entanglement with his/her surroundings into abyssal solipsistic egomania in line with the philosopher's other self-absorbed work.

51 Erik Swyngedouw and Henrik Ernstson, "Interrupting the Anthropo-obScene: Immuno-biopolitics and Depoliticizing Ontologies in the Anthropocene," *Theory, Culture & Society* 35, 6, 2018, 3–30, and Swyngedouw, op. cit.

52 Alain Brossat, *La Démocratie Immunitaire*, Paris: La Dispute 2003. Cf. Henrik Ernstson and Erik Swyngedouw, *Urban Political Ecology in the Anthropo-obscene: Interruptions and Possibilities*, London and New York: Routledge 2018.

53 Cf. Karolina Sobecka, "The Atmospheric Turn," in: Sigurd Bergmann and Forrest Clingerman (eds.), *Arts and Religion Responding to the Environment: Exploring Nature's Texture* (Studies in Environmental Humanities 6), Leiden: Brill Rodopi 2018, 43–58.

54 Cf. the widely distributed and debated article by David Wallace-Wells, "The Uninhabitable Earth: Famine, Economic Collapse, a Sun that Cooks Us: What Climate Change could Wreak – Sooner than you Think," *New York Magazine*, 9 July 2017, http://nymag.com/daily/intelligencer/2017/07/climate-change-earth-too-hot-for-humans.html, accessed 19 November 2017.

55 Irmgard Emmelhainz, "Images Do Not Show: The Desire to See in the Anthropocene," in: Heather Davis and Etienne Turpin (eds.), *Art in the Anthropocene: Encounters among Aesthetics, Politics, Environments and Epistemologies*, London: Open Humanities Press 2015, 131–142.

56 Rolita Machila, "Why Are Earth and God Angry?" in: Issue 20 (August 2008) in the Lutheran World Federation (LWF) "Thinking It Over" series, at https://www.lutheranworld.org/sites/default/files/DTS-Thinking-20.pdf.

57 Maria Antonaccio, "De-Moralizing and Re-Moralizing the Anthropocene," in: Deane-Drummond, Bergmann, and Vogt (eds.), op. cit., 121–137, 128.
58 Such as actor-network-theory, cosmological ethics of care, posthumanism, or new accelerationism.
59 Deane-Drummond, Bergmann, and Vogt, op. cit., 1–15, 2.
60 Franz Mauelshagen, "Redefining Historical Climatology in the Anthropocene," *The Anthropocene Review* 1, 2, 2014, 171–204, 172. Dipesh Chakrabarty, "Verändert der Klimawandel die Geschichtsschreibung?" *Transit. Europäische Revue* 41, 2011, 143–163.
61 Johann Gottfried Herder, *Outlines of a Philosophy of the History of Man*, London: Hansard 1800, 176 (orig. *Ideen zur Philosophie der Geschichte der Menschheit*, 4 volumes, Riga and Leipzig: Johann Friedrich Hartknoch 1784–1791, https://reader.digitale-sammlungen.de/resolve/display/bsb10897891.html, accessed 28 November 2019).
62 Cf. Sigurd Bergmann, "Religion at Work within Climatic Change: Eight Perceptions about Its Where and How," in: Deane-Drummond, Bergmann, and Vogt (eds.), op. cit., 67–84.
63 Cf., for example, the excellently differentiated analysis of the human signature on Earth's history by Waters et al., where "humans" and "humanity" throughout the whole investigation are simply treated in the singular and in general, probably due to the interest in geology and earth system analysis and the fact that the methods used are exclusively for *large-scale* exploration.
64 Cf. Jorge Otero-Pailos, "The Atmosphere as a Cultural Object," in: James Graham (ed.) with Caitlin Blanchfield, Alissa Anderson, Jordan H. Carver, and Jacob Moore (eds.), *Climates: Architecture and the Planetary Imaginary*, Zürich: Lars Müller Publishers 2016, 243–250.
65 Cf. Aleida Assmann, *Erinnerungsräume: Formen und Wandlungen des kulturellen Gedächtnisses*, München: C. H. Beck 1999; Jan Assmann, *Religion und kulturelles Gedächtnis: Zehn Studien*, München: Beck 2000/2006; Pierre Nora, "Between Memory and History: Les Lieux de Mémoire," *Representations* 26 (Special Issue: Memory and Counter-Memory), Spring 1989, 7–24.
66 One might therefore construct a new analytical method for exploring transdisciplinarily – with combined methods from climate impact science, geography, and environmental humanities – what I have called the "lived spaces of climatic change," that is, the specifically changing environments where processes of anthropogenic external impacts and local human adaptations, which again produce new kinds of impacts, can be studied in one common "Funktionskreis". S. Bergmann, unpubl., "Religion in Climatic Change," application to the European Research Council (ERC), March 2011.
67 Nicholas Maxwell, *From Knowledge to Wisdom: A Revolution in the Aims and Methods of Science*, Oxford: Blackwell 1984; "Can Scientific Method Help Us Create a Wiser World?" in: N. Dalal, A. Intezari, and M. Heitz (eds.), *Practical Wisdom in the Age of Technology: Insights, Issues and Questions for a New Millennium*, London: Gower 2016; http://discovery.ucl.ac.uk/1458254/, accessed 19 November 2017.
68 Cf. Celia E. Deane-Drummond, *Wonder and Wisdom: Conversations in Science, Spirituality and Theology*, London: Darton, Longman & Todd 2006.
69 The term "die armen Naturen" (the poor natures) was invented in the context of an emerging ecotheological liberation theology in Sigurd Bergmann, *Geist, der Natur befreit: Die trinitarische Kosmologie Gregors von Nazianz im Horizont einer ökologischen Theologie der Befreiung*, Mainz: Matthias-Grünewald Verlag 1995 (Engl. rev. ed. *CREATION SET FREE: The Spirit as Liberator of Nature*, with a Foreword by Jürgen Moltmann (Sacra Doctrina: Christian Theology for a Postmodern Age 4), Grand Rapids, MI, and Cambridge, UK: William B. Eerdmans Publishing Company 2005, Russian transl. Arkhangelsk: Arkhangelsk University Press 2003).

70 On spatial eschatology, see Vitor Westhelle, *Eschatology and Space: The Lost Dimension in Theology Past and Present*, New York: Palgrave Macmillan 2012, 121; Sigurd Bergmann, "Time Turned into Space – at Home on Earth: Wanderings in Eschatological Spatiality," in: S. Bergman (ed.), *Eschatology as Imagining the End: Faith between Hope and Despair*, London and New York: Routledge 2018, 88–112.

71 Cf. http://ipcc.ch/.

72 Cf. Ottmar Edenhofer and Michael Jakob, *Klimapolitik: Ziele, Konflikte, Lösungen*, München: C. H. Beck 2017, 118–123.

73 Harald Welzer, *Selbst Denken: Eine Anleitung zum Widerstand*, 7th edition, Frankfurt/M.: Fischer 2016, 133.

74 Leonard Cohen, in the album *The Future*, 1992.

75 For a deeper exploration of such questions, see Ernst M. Conradie, *Redeeming Sin? Social Diagnostics amid Ecological Destruction*, London: Lexington Books 2017. And for the radicality of remembering and continuing the Reformation today and tomorrow, see the comprehensive international project *Radicalized Reformation* and its thought-provoking seven volumes: www.radicalizing-reformation.com/index.php/en/theses.html, accessed 3 December 2019.

76 The Anthropocene narrative might also offer an analogy for some of the negative dimensions of the so-called New Cosmology and (grand) Story of the Universe as it has been critically investigated. Cf. Lisa H. Sideris, *Consecrating Science: Wonder, Knowledge, and the Natural World*, Oakland: University of California Press 2017.

77 Ernst M. Conradie, S. Bergmann, C. E. Deane-Drummond, and D. Edwards (eds.), *Christian Faith and the Earth: Current Paths and Emerging Horizons in Ecotheology*, London and New York: Bloomsbury 2014; Willis Jenkins, Mary Evelyn Tucker, and John Grim (eds.), *The Routledge Handbook of Religion & Ecology*, London and New York: Routledge 2017; John Hart (ed.), *The Wiley Blackwell Companion to Religion and Ecology*, Oxford: Wiley-Blackwell 2017; Laura Hobgood and Whitney Bauman (eds.), *The Bloomsbury Handbook of Religion and Nature*, London and New York Bloomsbury 2018; Ernst M. Conradie and Hilda P. Koster (eds.), *T&T Clark Handbook of Christian Theology and Climate Change*, New York: Bloomsbury 2019.

78 Johann-Albrecht Meylahn, "Doing Public Theology in the Anthropocene Towards Life-Creating Theology," *Verbum et Ecclesia* 36, July 2015. While Meylahn pleads for us "to seek God in the text, or rather as part of the grammatisation" and to do contextual theology about God in the Anthropocene (rather than about a God beyond), my approach here implies both doing this and encountering God in the not-yet-seen future and space beyond the age of the humans.

79 Marion Grau, "The Revelations of Global Climate Change: A Petro-Theology," in: Bergmann (ed.), *Eschatology as Imagining the End*, op. cit., 45–60.

80 For the notion of "technocene", see Alf Hornborg, "The Political Ecology of the Technocene: Uncovering Ecologically Unequal Exchange in the World-System," in: C. Hamilton, Ch. Bonneuil, and F. Gemenne (eds.), *The Anthropocene and the Global Environmental Crisis: Rethinking Modernity in a New Epoch*, London: Routledge 2015, 57–69. For the notion of "technical space", see Sigurd Bergmann and Tore Sager (eds.), *The Ethics of Mobilities: Rethinking Place, Exclusion, Freedom and Environment*, London: Routledge 2008.

81 Keller does not tumble into this abyss but offers quite a homemade – mixed up with Pope Francis' concept of an "integral ecology" (in his *Laudato Si´*) – understanding of the Ecocene as a kind of a subject, an "earthome" (earth+home) that "at once warns and invites" and that "is to be nourished if it is to nourish us". Catherine Keller, *Political Theology of the Earth: Our Planetary Emergency and the Struggle*, New York: Columbia University Press 2018, 92.

82 Arjen E. J. Wals, Joseph Weakland, and Peter Blaze Corcoran, "Preparing for the Eco-cene: Envisioning Futures for Environmental and Sustainability Education," *Japanese Journal of Environmental Education* 26, 4, 2017, 71–76, 72.
83 Rachel Armstrong, "Architecture for the Ecocene," 31 July 2015, https://architecture ukraine.org/rachel-armstrong-architecture-for-the-ecocene/, accessed 20 November 2017.
    Cf. Joanna Boehnert, *Design, Ecology, Politics: Towards the Ecocene*, London and New York: Bloomsbury 2018. Cf. Joanna Boehnert, "Naming the Epoch: Anthropocene, Capitalocene, Ecocene," https://ecolabsblog.com/2016/05/22/naming-the-epoch-anthropocene-capitalocene-ecocene/, accessed 20 November 2017.
84 Robert Steiner, "From Anthropocene to Ecocene by 2050?" 23 October 2017: www.huffingtonpost.com/entry/from-anthropocene-to-ecocene-by-2050_us_59e7b66ce4b0e60c4aa3678c, accessed 20 November 2017. Thomas Berry had earlier coined the term *ecozoic era* to overcome anthropocentrism, but the notion of an Ecocene includes not only living beings but also all planetary forms.
85 On the significance of "shared future" in the context of "making oneself at home in the future", see Scott and Rodwell in their introductory chapter "Dialogues of Place and Belonging," in: John Rodwell and Peter Manley Scott (eds.), *At Home in the Future: Place & Belonging in a Changing Europe* (Studies in Religion and the Environment 11), Berlin: LIT 2016, 9.
86 Aleida Assmann, *Ist die Zeit aus den Fugen? Aufstieg und Fall des Zeitregimes der Moderne*, München: Hanser 2013, 247.
87 Cf. Westhelle, op. cit.
88 Robert Vosloo, "Time Out of Joint and Future-Oriented Memory: Engaging Dietrich Bonhoeffer in the Search for a Way to Deal Responsibly with the Ghosts of the Past," *Religions* 8, 42, 2017.
89 For an environmental ethic in the frame of discourse ethics, see Konrad Ott, *Umweltethik zur Einführung*, 2nd edition, Hamburg: Junius 2014; and for a differentiated theological approach to such an environmental discourse ethics, see Christ of Hardmeier and Konrad Ott, *Naturethik und biblische Schöpfungserzählung: Ein diskurstheoretischer und narrativ-hermeneutischer Brückenschlag* (*Nature Ethics and Biblical Creation Story: Bridging the Gap through Discourse Theory and Narrative Hermeneutics*), Stuttgart: Kohlhammer 2015.
    For a discussion of the pros and cons of an environmental ethics of responsibility versus an ecological discourse ethics, see Sigurd Bergmann, "Diskursiv bioetik – för offrens skull," in: Uno Svedin and Anne-Marie Thunberg (eds.), *Miljöetik – för ett samhälle på människans och naturens villkor* (FRN Rapport 94:2), Stockholm: Miljödepartementet/ Forskningsrådsnämnden 1994, 68–89, ["Diskursive Bioethik – um der Opfer willen," in: S. Bergmann, *Geist, der lebendig macht: Lavierungen zur ökologischen Befreiungstheologie*, Frankfurt/M.: IKO 1997].
    An environmental ethics and an ethics of responsibility need not, of course, contradict each other; both could be connected in a statement that might follow Haber, Held, and Vogt's plea for a new culture of responsibility ("eine neue Kultur der Verantwortung," p. 13) that necessarily includes the qualities of a non-violent dialogue as suggested in discourse ethics. Wolfgang Haber, Martin Held, and Markus Vogt (eds.), *Die Welt im Anthropozän: Erkundungen im Spannungsfeld zwischen Ökologie und Humanität*, München: Oekom 2016.
90 Such as the ethics of care, African Ubuntu ethics, or indigenous ethics where the land, the ocean, the forest, etc. are moral subjects rather than objects that can place the human and the community under obligation.
91 On the exciting history of elaborating "rights of nature", cf. Goethe's scientific and poetic reflections, the discourse of lawyers and theologians and their manifesto in the 1980s, the Nordic discussion in the sociology of law that caused the establishing of an environmental court in Finland, and the most recent decisions of Ecuador (2008) and Bolivia (2010) to pass national laws giving nature constitutional rights.

Johann Wolfgang Goethe, "Recht und Pflicht. Zur Naturwissenschaft II. 2.," in: *Zur Naturwissenschaft überhaupt, besonders zur Morphologie: Erfahrung, Betrachtung, Folgerung durch Lebensereignisse verbunden*, Münchner Ausgabe Band 12, Munich 1989, 772; S. Bergmann, "Naturens värde och människans vördnad: Varför man bör tillerkänna naturen dess rättigheter," *Retfærd* 65, 1994, 76–86 [Nature's Value and the Human's Dignity: Why one shall Award Nature its Rights]; http://pdba.georgetown.edu/Constitutions/Ecuador/english08.html, https://therightsofnature.org/wp-content/uploads/pdfs/FINAL-UNIVERSAL-DECLARATION-OF-THE-RIGHTS-OF-MOTHER-EARTH-APRIL-22-2010.pdf, accessed 2 December 2019.

92 Cobb describes theology as "self-conscious Christian reflection about important matters". John B. Cobb. Jr., "The Role of Theology of Nature in the Church," in: Charles Birch, William Eakin, and Jay B. McDaniel (eds.), *Liberating Life: Contemporary: Approaches to Ecological Theology*, Maryknoll: Orbis 1990, 261–272, 262.

93 Leonard Cohen launched the album *Popular Problems* on his 70th birthday (Columbia 2014).

94 Dietrich Bonhoeffer, "Die Kirche ist nur Kirche, wenn sie für andere da ist," in: *Widerstand und Ergebung*, DBW, Vol. 8, *München: Random House 2011,* 560.

95 Bronislaw Szerszynski, "Life in the Open Air," chapter 3 in: Dirk Evers, Michael Fuller, Antje Jackelén, and Knut-Willy Sæther (eds.), *Issues in Science and Theology: What Is Life?* Cham: Springer International 2016, 27–41.

96 Szerszynski, op. cit., 33.

97 Bron Szerszynski, "Reading and Writing the Weather: Climate Technics and the Moment of Responsibility," *Theory, Culture & Society* 27, 2010, 9–30, 21.

98 Lisa Robertson, "The Weather: A Report on Sincerity," Washington, DC: Poetry 2001 Anthology, www.dcpoetry.com/anthology/242, accessed 17 October 2018.

99 Ibid.

100 C. Donald Ahrens, *Meteorology Today: An Introduction to Weather, Climate, and the Environment*, 6th edition, Pacific Grove: Brooks/Cole 2000, 15; and Helmut Kraus, *Die Atmosphäre der Erde*, 4th edition, Berlin, Heidelberg, and New York: Springer 2004, 11.

101 Tim Ingold, "Lines and the Weather," The Daphne Mayo Lecture, presented at the University of Queensland Art Museum on Wednesday 16 October 2013, p. 12.

102 Sigurd Bergmann, "Atmospheres of Synergy: Towards an Eco-Theological Aesth/Ethics of Space," *Ecotheology: The Journal of Religion, Nature and the Environment* 11, 3, 2006, 327–357.

103 Otto Bollnow, *Mensch und Raum*, 10th edition, Stuttgart: Kohlhammer 2004 (1963), 13.

104 Gernot Böhme, *Für eine ökologische Naturästhetik*, Frankfurt am Main: Suhrkamp 1989, 8. Cf. J.D. Porteous, *Environmental Aesthetics: Ideas, Politics and Planning*, London and New York: Routledge 1996, 5–41.

105 Gernot Böhme, *Atmosphäre: Essays zur neuen Ästhetik*, Frankfurt am Main: Suhrkamp 1995, 23.

106 On atmospheric spaces, see Hermann Schmitz, *Atmosphären*, Freiburg and Munich: Alber 2014; Gernot Böhme, *Aisthetik: Vorlesungen über Ästhetik als allgemeine Wahrnehmungslehre*, Munich: Fink 2001; Tonino Griffero, *Atmospheres: Aesthetics of Emotional Spaces*, Farnham: Ashgate 2014.

107 Griffero, op. cit.

108 Mădălina Diaconu, "City Walks and Tactile Experience," *Contemporary Aesthetics* 9, 2011, 10: https://contempaesthetics.org/newvolume/pages/article.php?articleID=607, accessed 3 December 2019.

109 Gernot Böhme, "Atmosphere as the Fundamental Concept of a New Aesthetics," *Thesis Eleven* 36, 1993, 113–126, 119.

110 "Eine Atmosphäre im hier gemeinten Sinn ist eine randlose, unteilbar ausgedehnte Besetzung eines flächenlosen Raumes". Hermann Schmitz, "Von der Scham zum Neid," in: Michael Großheim, Anja Kathrin Hild, Corinna Lagemann, and Nina

Trčka (eds.), *Leib, Ort, Gefühl: Perspektiven der räumlichen Erfahrung*, Freiburg and München: Karl Alber 2015, 19–34, 20.

111 Kerstin Andermann and Undine Eberlein, "Einleitung: Gefühle als Atmosphären? Die Provokation der Neuen Phänomenologie," in: Kerstin Andermann and Undine Eberlein (eds.), *Gefühle als Atmosphären: Neue Phänomenologie und philosophische Emotionstheorie* (Deutsche Zeitschrift für Philosophie, Sonderband 29), Berlin: De Gruyter Akademie Forschung 2011, 7–17, 10. Emotions are for Schmitz, 21, "räumlich ergossene, leiblich ergreifende Atmosphären". Hermann Schmitz, "Entseelung der Gefühle," in: Andermann and Eberlein (eds.), op. cit., 21–33.

112 Mónika Dánél, "Atmospheric Adaptation as Cultural Translation (Ádám Bodor – Gábor Ferenczi: *The Possibilities of Making Friends*, 2007)," *Contact Zones: Studies in Central and Eastern European Film and Literature: A Biannual Online Journal* 2016/2, 6–23, http://contactzones.elte.hu/atmospheric-adaptation, accessed 3 December 2019.

113 Tim Ingold, *Lines: A Brief History*, London: Routledge 2007,18f.

114 Schmitz, according to Catherine Newmark, *Der philosophische Wetterbericht (3): Was das Wetter mit unseren Gefühlen zu tun hat*, www.deutschlandfunkkultur.de/der-philosophische-wetterbericht-3-was-das-wetter-mit.2162.de.html?dram:article_id=328438, accessed 27 October 2018.
     Cf. also Gernot Böhme, "Das Wetter und die Gefühle: Für eine Phänomenologie des Wetters," in: Andermann and Eberlein (eds.), op. cit., 153–166.

115 Böhme, "Das Wetter und die Gefühle," op. cit.

116 Ingold, op. cit., 20.

117 Ibid., 20.

118 Ibid.

119 On the psychology of affective atmosphere, see Ben Anderson, "Affective Atmospheres," *Emotion, Space and Society* 2, 2, 2009, 77–81.

120 Mădălina Diaconu, "Longing for Clouds – Does Beautiful Weather Have to Be Fine?" *Contemporary Aesthetics* 13, 2015, 1, https://digitalcommons.risd.edu/liberalarts_contempaesthetics/vol13/iss1/16/, accessed 3 December 2019.

121 Ibid., 2.

122 Ibid., 4.

123 Saint Ephrem, *Hymns on Paradise*, Crestwood, New York: St Vladimir's Seminary Press 1990, 149, quoted according to Diaconu, "Longing for Clouds," op. cit., 5.

124 Master painters in art history have in such a sense already documented climatic change in their atmospheric landscape painting. Cf. C.S. Zerefos, P. Tetsis, A. Kazantzidis, V. Amiridis, S.C. Zerefos, J. Luterbacher, K. Eleftheratos, E. Gerasopoulos, S. Kazadzis, and A. Papayannis, "Further Evidence of Important Environmental Information Content in Red-to-Green Ratios as Depicted in Paintings by Great Masters," *Atmospheric Chemistry and Physics* 14, 2014, 2987–3015, www.atmos-chem-phys.net/14/2987/2014/, doi: 10.5194/acp-14-2987-2014.

125 Mădălina Diaconu, "Wetter, Welten, Wirkungen: Sinnverschiebungen der Atmosphäre," in: Christiane Heibach (ed.), *Atmosphären: Dimensionen eines diffusen Phänomens*, München: Wilhelm Fink 2012, 84–99, 85.

126 Diaconu, op. cit., 87.

127 Diaconu refers to Roth's dissertation as evidence, who also includes notes on the sky's beauty. But one can put forward as further evidence the national German TV channel ZDF's practice of decorating each evening's main weather forecast with one viewer's amateur photo, often following a classical naturalistic aesthetics of nature and the landscape. In addition, they organise the "Wetterfoto der Woche" (the week's weather photo): www.zdf.de/nachrichten/zdf-morgenmagazin/wetterfoto-der-woche-100.html, accessed 29 November 2019. The Swedish National TV (SVT) has for years now published weather pictures online weekly from their viewers' amateur photographers. www.svt.se/nyheter/amne/Tittarnas_v%C3%A4derbilder_till_SVT_V%C3%A4der,

accessed 29 November 2019. It has become a popular activity to complement scientific meteorological investigations with photos from citizens' ordinary life worlds. Related to such fruitful participative activities, the Swedish Meteorological Service (SMHI) invites citizens to receive training and serve as volunteer "climate observers", assisting in collecting data from selected places. www.smhi.se/jobba-pa-smhi/jobba-pa-smhi/bli-var-nya-klimatobservator-1.121564, accessed 29 November 2019. Regarded in the lens of what Roni Horn, in note 7 above, made us aware of, such broad and diversified archives of *weather reporting us* might offer a valuable, efficient method for bridging the gap between meteorological and aesthetic atmospheric thinking, and for enhancing the sociocommunicative process, with the potential to catalyse multifaceted weather wisdom.

128 Diaconu (my transl.), op. cit., 125. Cf. on *the strange(r)* Sigurd Bergmann, "The Strange and the Self: Visual Arts and Theology in Aboriginal and Other (Post-)Colonial Spaces," in: Oleg V. Bychkov and James Fodor (eds.), *Theological Aesthetics After Von Balthasar*, Aldershot: Ashgate 2008, 201–223.

129 Cf. also on this topic the remarkably inspiring work of environmental artists Reiko Goto and Timothy Collins on empathy with nature. Reiko Goto and Tim Collins, "The Black Wood: Relations, Empathy and a Feeling of Oneness in Caledonian Pine Forests," in: Sigurd Bergmann and Forrest Clingerman (eds.), *Arts, Religion and the Environment: Exploring Nature's Texture* (Studies in Environmental Humanities 6), Amsterdam: Brill/Rodopi 2018, 117–148.

Accordingly, one might also pay heed to what Braungart emphasises in his interpretation of weather in modern German nature lyrics, namely that we, in the context of climatology, can obviously perceive a new and highly relevant interrelation between science and poetry, whereby "actual weather knowledge is regarded anew as existentially significant". Braungart, op. cit., 303.

130 In a rich and very valuable recent publication, which unfortunately first reached me just before finalising the manuscript of this book, Büttner and Braungart have edited important contributions to such a synthetic study in their work on the nature-culture-relation in the frame of "wind and weather". The editors, p. 11, aptly make us aware not only of the diversity of practices and knowledge forms with regard to the atmosphere but also of their regional and historical differentiation. Braungart and Büttner (eds.), op. cit.

131 Not a "first" and "second nature" as in Walther Benjamin's influential thinking but a complex total nature to be approached from all angles. An inspiring example of such an approach can be visited and experienced in the *Dessau-Wörlitzer Gartenreich*, where the landscape park, established in the late 19th century under Fürst Leopold III, presents itself as a *Gesamtkunstwerk*. www.gartenreich.de/de/, accessed 29 November 2019. That weather plays a crucial role in such a place is obvious, while it is still obscure how both the original designers and the contemporary visitors experience(d) and reflect(ed) it. Such a place would probably be the right one for mining more deeply what this chapter has in mind.

132 The term "New materialism" was coined by Manuel DeLanda and Rosi Braidotti in the second half of the 1990s. It opposes dichotomist thinking and searches for new ways out of these traditions. Nature and culture are always understood as *naturecultures* (D. Haraway) and the human mind is always regarded as material. In Karen Barad's view, for example, quantum physics inspires a new way of entangling matter and meaning, in a way that she has circumscribed as "diffractive methodology, a method of diffractively reading insights through one another, building new insights, and attentively and carefully reading for differences that matter in their fine details, together with the recognition that there intrinsic to this analysis is an ethics that is not predicated on externality but rather entanglement". https://quod.lib.umich.edu/o/ohp/11515701.0001.001/1:4.3/-new-materialism-interviews-cartographies?rgn=div2;view=fulltext, accessed 29 November 2019.

For a survey, see the EU-supported project "New Materialism – Networking European Scholarship on 'How Matter Comes to Matter'": https://newmaterialism.eu/, accessed 29 November 2019.

133 Catherine Keller, "Tingles of Matter, Tangles of Theology," in: Catherine Keller and Mary-Jane Rubenstein (eds.), *Entangled Worlds: Religion, Science, and New Materialisms*, New York: Fordham University Press 2017.

134 On the distinction of "power over" and "power with" others, see Karen J. Warren, "A Feminist Philosophical Perspective on Ecofeminist Spiritualities," in: Carol J. Adams (ed.), *Ecofeminism and the Sacred*, New York: Continuum 1993, 119–132, 122f.

The German term *Abblendung* [dimming] was central in the post-war German critical thinkers' struggle with our common past (which the German-born author is also a part of). Cf. Theodor W. Adorno, *Minima Moralia: Reflexionen aus dem beschädigten Leben*, Berlin and Frankfurt am Main: Suhrkamp 1950. The term also became important in Adorno's aesthetic theory. Cf. Sigurd Bergmann, "Trinitarian Cosmology in God's Liberating Movement: Exploring some Signature Tunes in the *Opera of Ecologic Salvation*," *Worldviews* 14, 2010, 185–205, 185f.

135 Whitney A. Bauman, "Comparative Methods in Spatial Approaches to Religion," in: S. Bergmann (ed.), *Spatial Turns in Religion and the Environment*, Special Issue of *Worldviews: Global Religions, Culture, and Ecology* 20, 3, 2016, 311–322.

136 Catherine Keller invented the notion of "inter-carnation" as being-members-of-each-other. Catherine Keller, "Members of Each Other: Intercarnation, Gender and Political Theology," Workshop presentation at the Faculty of Theology in Lund, 16 September 2019. Cf. Catherine Keller, *Intercarnations: Exercises in Theological Possibility*, New York: Fordham University Press 2017.

137 This is explicitly shown by Lüdicke in the impressive line she has drawn from the early to the contemporary history of meteorological investigation of the higher atmosphere. Cornelia Lüdecke, "Vom Berg zur freien Atmosphäre: Die Erforschung der dritten Dimension seit dem 17. Jahrhundert," in: Braungart and Büttner (eds.), op. cit., 357–373.

138 Cf. Diaconu, "Longing for Clouds," op. cit., 12.

139 Deborah R. Coen, "Seeing Planetary Change: Down to the Smallest Wildflower," in: Blanchfield, Anderson, Carver, and Moore (eds.), op. cit., 34–39, 39.

140 Cf. Celia Deane-Drummond, *Wonder and Wisdom: Conversations in Science, Spirituality, and Theology*, West Conshohocken: Templeton Press 2006.

141 Johann Wolfgang Goethe, *Bedeutende Fördernis durch ein einziges geistreiches Wort*, 1823, 306: "Der Mensch kennt nur sich selbst, insofern er die Welt kennt, die er nur in sich und sich nur in ihr gewahr wird," in: *Zur Naturwissenschaft überhaupt, besonders zur Morphologie: Erfahrung, Betrachtung, Folgerung, durch Lebensereignisse verbunden*, München: Hanser 1989 (1817–24), 306–309.

142 Cf. Kant, who assumes that modern science has vainly ignored the question of wisdom: "Nach Weisheit frägt niemand, weil sie die Wissenschaft, die ein Werkzeug der Eitelkeit ist, sehr ins Enge bringt" (*Reflections* 16:66, Leipzig 1882). Nevertheless the enlightened philosophical giant can also circumscribe his own work as "die 'Kritik und Vorschrift der gesammten Weltweisheit als eines Ganzen', kurz die 'Kritik der Vernunft'" (the 'Critique and Instruction of the total World-wisdom as a whole', in short, the 'Critique of Reason'). *Reflexionen Kants zur kritischen Philosophie: aus Kants handschriftlichen Aufzeichnungen*, ed. by Benno Erdmann, Leipzig: Fues's Verlag (R. Reisland) 1882–84, XXI, https://archive.org/stream/reflexionenkants-00kantuoft/reflexionenkants00kantuoft_djvu.txt, accessed 3 December 2019.

Index